Recent Titles in This Series

182 **Selman Akbulut, Editor,** Real algebraic geometry and topology, 1995
181 **Mila Cenkl and Haynes Miller, Editors,** The Čech Centennial, 1995
180 **David E. Keyes and Jinchao Xu, Editors,** Domain decomposition methods in scientific and engineering computing, 1994
179 **Yoshiaki Maeda, Hideki Omoro, and Alan Weinstein, Editors,** Symplectic geometry and quantization, 1994
178 **Hélène Barcelo and Gil Kalai, Editors,** Jerusalem Combinatorics '93, 1994
177 **Simon Gindikin, Roe Goodman, Frederick P. Greenleaf, and Paul J. Sally, Jr., Editors,** Representation theory and analysis on homogeneous spaces, 1994
176 **David Ballard,** Foundational aspects of "non"standard mathematics, 1994
175 **Paul J. Sally, Jr., Moshe Flato, James Lepowsky, Nicolai Reshetikhin, and Gregg J. Zuckerman, Editors,** Mathematical aspects of conformal and topological field theories and quantum groups, 1994
174 **Nancy Childress and John W. Jones, Editors,** Arithmetic geometry, 1994
173 **Robert Brooks, Carolyn Gordon, and Peter Perry, Editors,** Geometry of the spectrum, 1994
172 **Peter E. Kloeden and Kenneth J. Palmer, Editors,** Chaotic numerics, 1994
171 **Rüdiger Göbel, Paul Hill, and Wolfgang Liebert, Editors,** Abelian group theory and related topics, 1994
170 **John K. Beem and Krishan L. Duggal, Editors,** Differential geometry and mathematical physics, 1994
169 **William Abikoff, Joan S. Birman, and Kathryn Kuiken, Editors,** The mathematical legacy of Wilhelm Magnus, 1994
168 **Gary L. Mullen and Peter Jau-Shyong Shiue, Editors,** Finite fields: Theory, applications, and algorithms, 1994
167 **Robert S. Doran, Editor,** C^*-algebras: 1943–1993, 1994
166 **George E. Andrews, David M. Bressoud, and L. Alayne Parson, Editors,** The Rademacher legacy to mathematics, 1994
165 **Barry Mazur and Glenn Stevens, Editors,** p-adic monodromy and the Birch and Swinnerton-Dyer conjecture, 1994
164 **Cameron Gordon, Yoav Moriah, and Bronislaw Wajnryb, Editors,** Geometric topology, 1994
163 **Zhong-Ci Shi and Chung-Chun Yang, Editors,** Computational mathematics in China, 1994
162 **Ciro Ciliberto, E. Laura Livorni, and Andrew J. Sommese, Editors,** Classification of algebraic varieties, 1994
161 **Paul A. Schweitzer, S. J., Steven Hurder, Nathan Moreira dos Santos, and José Luis Arraut, Editors,** Differential topology, foliations, and group actions, 1994
160 **Niky Kamran and Peter J. Olver, Editors,** Lie algebras, cohomology, and new applications to quantum mechanics, 1994
159 **William J. Heinzer, Craig L. Huneke, and Judith D. Sally, Editors,** Commutative algebra: Syzygies, multiplicities, and birational algebra, 1994
158 **Eric M. Friedlander and Mark E. Mahowald, Editors,** Topology and representation theory, 1994
157 **Alfio Quarteroni, Jacques Periaux, Yuri A. Kuznetsov, and Olof B. Widlund, Editors,** Domain decomposition methods in science and engineering, 1994
156 **Steven R. Givant,** The structure of relation algebras generated by relativizations, 1994
155 **William B. Jacob, Tsit-Yuen Lam, and Robert O. Robson, Editors,** Recent advances in real algebraic geometry and quadratic forms, 1994

(*Continued in the back of this publication*)

CONTEMPORARY MATHEMATICS

182

Real Algebraic Geometry and Topology

A Conference on Real Algebraic Geometry and Topology
December 17–21, 1993
Michigan State University

Selman Akbulut
Editor

American Mathematical Society
Providence, Rhode Island

A conference on Real Algebraic Geometry and Topology was held at Michigan State University, East Lansing, Michigan, December 17–21, 1993.

1991 *Mathematics Subject Classification*. Primary 14C25, 57R19, 15P05.

Library of Congress Cataloging-in-Publication Data

Conference on Real Algebraic Geometry and Topology (1993: Michigan State University)
Real algebraic geometry and topology: a Conference on Real Algebraic Geometry and Topology, December 17–21, 1993, Michigan State University / Selman Akbulut, editor.
p. cm. — (Contemporary mathematics, ISSN 0271-4132; v. 182)
Includes bibliographical references and index.
ISBN 0-8218-0292-5
1. Geometry, Algebraic—Congresses. 2. Algebraic topology—Congresses. I. Akbulut, Selman, 1949– . II. Title. III. Series: Contemporary mathematics (American Mathematical Society); v. 182.
QA564.C664 1993
516.3′5—dc20 94-44468
CIP

Printed in the United States of America.

The paper used in this book is acid-free and falls within the guidelines
established to ensure permanence and durability.
Printed on recycled paper.
All articles in this volume were printed from copy prepared by the authors.
Some articles were typeset using AMS-TEX or AMS-LATEX,
the American Mathematical Society's TEX macro systems.

10 9 8 7 6 5 4 3 2 1 00 99 98 97 96 95

Contents

PREFACE

The Michigan State University Conference on Real Algebraic Geometry and Topology was held at the MSU campus on December 17–21, 1993. It gave many experts a chance to review recent results as well as to discuss a wide variety of new ideas. Speakers included E. Bierstone, M. Buchner, A. Degtyarev, H. Dovermann, A. Gabrielov, J. Havlicek, J. Huisman, I. Itenberg, K. Kurdyka, H. King, G. Mikhalkin, P. Milman, Y. Ozan, R. Silhol, F. Sottile, D. Trotman, and O. Viro. Topics included resolution theorems, nonsingular algebraic sets and algebraic homology, topology of singular real algebraic sets, distribution of real algebraic sets in projective space, and moduli spaces of algebraic structures.

We would like to thank the N.S.F., the Institute for Mathematics and its Applications, and the Michigan State University Foundation for supporting this conference, and the AMS for agreeing to publish these Proceedings. We also want to thank the MSU math department in general and Professors Dick Phillips, Barbara Miller, Sterling Tryon-Hartwig, and Kathy Trebilcott in particular, and thanks especially to Mary Reynolds for coordinating day-to-day workings of the conference. Finally, many thanks to Professor Craig Huneke for prompt editing, and to Donna Harmon of the AMS for the faultless organization of these Proceedings.

Contemporary Mathematics
Volume **182**, 1995

Vanishing of Characteristic Classes of a Real Algebraic Variety

ALEXANDER DEGTYAREV

ABSTRACT. A relationship between the characteristic classes of a maximal real algebraic variety and those of its complexification is obtained, which allows to find the condition for existence of canonical additional structures on the real part coming from the complexification, to obtain conditions for existence of equivariant structures on the complexification, and may have other applications.

1. Introduction

1.1. The purpose of this paper is to study the relationship between the characteristic classes (or, more precisely, their direct images in some auxiliary spaces) of a maximal real algebraic variety and those of its complexification. (A *maximal* real algebraic variety, or, briefly, just M-variety, is a real algebraic variety A whose sets $\mathbb{R}A$ and $\mathbb{C}A$ of, respectively, real and complex points represent the following extremal case of the Smith-Thom inequality: $b_*(\mathbb{R}A) = b_*(\mathbb{C}A)$, where $b_*(\cdot)$ denotes the total Betti number with $\mathbb{Z}_2$-coefficients. These objects traditionally play a special rôle in the topology of real algebraic varieties: it turns out that there are certain rather strict restrictions to the topology of their real parts $\mathbb{R}A$, which gradually disappear when the total Betti number decreases.) The main result, Theorem 2.3 and Corollary 2.4, has been inspired by and is a direct generalization of the following result, which is due to V. Arnol′d, V. Kharlamov, and V. Rokhlin:

1991 *Mathematics Subject Classification.* Primary 57S17; Secondary 14P25.

Key words and phrases. Characteristic class, real algebraic variety, involution.

This work was partially supported by Scientific and Technical Council of Turkey and Middle East Technical University (Ankara)

THEOREM (see [**K**], [**N**], or [**R**]). *Let A be a smooth projective real algebraic M-variety of dimension n, and let $\mathbb{C}A$ and $\mathbb{R}A$ be its sets of complex and real points, respectively. Then the fundamental class $[\mathbb{R}A]$ is Poincaré dual in $H_n(\mathbb{C}A;\mathbb{Z}_2)$ to the Wu class $v_n(\mathbb{C}A)$.*

1.2. Motivation. Originally, this work was motivated by the following problem: Given an n-dimensional real algebraic variety A, one can define certain *higher inclusion maps,* or *Viro homomorphisms,* $e_p^r\colon H_p(\mathbb{R}A;\mathbb{Z}_2) \dashrightarrow H_{p+r}(\mathbb{C}A;\mathbb{Z}_2)$ (see, e.g., Degtyarev [**D**]; the dashed arrow means that these are partial homomorphisms, i.e., homomorphisms defined on a subgroup of $H_p(\mathbb{R}A;\mathbb{Z}_2)$ modulo a subgroup of $H_{p+r}(\mathbb{C}A;\mathbb{Z}_2)$). If, for some $r \leqslant n$, $e_n^0[\mathbb{R}A] = e_n^1[\mathbb{R}A] = \dots = e_n^r[\mathbb{R}A] = 0$, then one can define some canonical partial Stiefel orientations of dimensions $n-1, n-2, \dots, n-r-1$ on $\mathbb{R}A$ (this is a generalization of Rokhlin's construction of the complex orientation on a separating real algebraic curve; see Viro [**V1**], [**V2**] and Degtyarev [**D**] for details). Thus, it is interesting to know when the above condition is satisfied, at least if A is an M-variety. If $r = 0$, the answer is given by the theorem cited above. If $r > 0$, the condition can be reformulated in terms of the spaces $\mathcal{S}^r\mathbb{C}A$ considered in this paper: it is equivalent to vanishing of the direct image of $1 \in H^0(\mathcal{S}^r\mathbb{R}A;\mathbb{Z}_2)$ in $H^{n+r}(\mathcal{S}^r\mathbb{C}A;\mathbb{Z}_2)$. Though Theorem 2.3 and Corollary 2.4 do not give a general answer to this question, under certain circumstances they do provide a sufficient condition for the existence of the canonical Stiefel orientations (see the remark after Corollary 2.4).

Among other possible applications of the results of this paper I would like to mention the following:

(1) The characteristic classes of $\mathbb{R}A \times \mathbb{R}\mathrm{p}^r$ whose images in $H^*(\mathcal{S}^r\mathbb{C}A;\mathbb{Z}_2)$ vanish according to Corollary 2.4 give rise to canonical structures on $\mathbb{R}A$ which are different from Stiefel orientations;
(2) Theorem 2.3 gives some restrictions to the topology of manifolds which admit M-involutions (see 2.2 for the definition);
(3) Relations between the characteristic classes of $\mathbb{R}A \times \mathbb{R}\mathrm{p}^k$ and those of $\mathcal{S}^j\mathbb{C}A$ give, under certain circumstances, necessary and sufficient conditions for the existence of equivariant structures on $\mathbb{C}A$.

I intend to devote a separate publication to these and other applications of the results obtained in the paper.

Acknowledgments. This work was partially completed during my stay in Middle East Technical University, Turkey. I would like to express my profound gratitude to the members of the Mathematical Department for their hospitality and support. I am also grateful to V. Kharlamov, who introduced me to the history of the subject.

2. Main result

2.1. Notation. Let us fix a paracompact topological space X with involution $c\colon X \to X$. Let $F = \operatorname{Fix} c$ be the fixed point set of c, and let $\operatorname{in}\colon F \hookrightarrow X$ be the inclusion. Denote by $\mathcal{S}^k_c X$, or just $\mathcal{S}^k X$, $0 \leqslant k \leqslant \infty$, the twisted product

$$\mathcal{S}^k X = X \times_{\mathbb{Z}_2} S^k = X \times S^k / \{(x, \mathbf{t}) \sim (cx, -\mathbf{t})\},$$

where $S^k \in \mathbb{R}^{k+1}$ is considered with the standard antipodal involution $\mathbf{t} \mapsto -\mathbf{t}$. Since $\mathbb{Z}_2$ is an abelian group, $\mathcal{S}^k X$ is also a $\mathbb{Z}_2$-space.

The standard inclusions $S^k \hookrightarrow S^l$, $k \leqslant l$, induce equivariant embeddings $j_{kl} \colon \mathcal{S}^k X \hookrightarrow \mathcal{S}^l X$. Furthermore, there are canonical homeomorphisms $\mathcal{S}^0 X = X$ and $\mathcal{S}^k F = F \times \mathbb{R}\mathrm{p}^k$. Thus, we have the following canonical diagram:

$$\begin{array}{ccccc} F \times \mathbb{R}\mathrm{p}^k = \mathcal{S}^k F & \xhookrightarrow{\mathrm{in}_k} & \mathcal{S}^k X & \xhookleftarrow{j_k} & X = \mathcal{S}^0 X, \\ & & \downarrow & & \\ & & \mathbb{R}\mathrm{p}^k = \mathcal{S}^k \mathrm{pt} & & \end{array}$$

where we let $j_k = j_{0k}$. In particular, we see that $H^*(\mathcal{S}^k F) = H^*(F) \otimes H^*(\mathbb{R}\mathrm{p}^k)$, and both $H^*(\mathcal{S}^k X)$ and $H^*(\mathcal{S}^k F)$ are algebras over $H^*(\mathbb{R}\mathrm{p}^k) = \mathbb{Z}_2[h]/h^{k+1}$. (Throughout the paper the coefficients of all the homology and cohomology groups are $\mathbb{Z}_2$.)

2.2. Given a smooth submanifold F of a smooth manifold X, we denote by τF and νF the tangent bundle of F and normal bundle of F in X respectively. Let us also denote by η the tautological linear bundle over $\mathbb{R}\mathrm{p}^k$, and by $[\alpha]_r$ the r-dimensional component of a (non-homogeneous) class α.

We say that c is an *M-involution* on X, or that (X, c) is an *M-manifold* with involution, if $b_*(F) = b_*(X)$, where $b_*(\cdot) = \sum_{i=0}^{\infty} b_i(\cdot)$ is the total Betti number (with $\mathbb{Z}_2$-coefficients). Note that this is an extremal case of the following Smith-Thom inequality: $b_*(F) \leqslant b_*(X)$.

2.3. MAIN THEOREM. *Let c be a smooth M-involution on a smooth manifold X, $\dim X = N$. Then for any pair of integers q, r such that $0 \leqslant q \leqslant r \leqslant N/2$ one has*

$$\left[(\mathrm{in}_N)_! V \cdot \mathrm{Sq}^{-1}\{h^{2q}(1+h)^N\}\right]_{N+r} = (j_N)_! v_r(X),$$

where $V = v(\tau F) v^{-1}(\nu F \otimes \eta) \in H^*(F \times \mathbb{R}\mathrm{p}^N)$ and $v(\,\cdot\,)$ is the total Wu class of a vector bundle.

PROOF is given in 6.6.

REMARK. Here and in several other places the Gysin homomorphism ($(\mathrm{in}_N)_!$ in this case) may be applied to either V or $V \cdot \mathrm{Sq}^{-1}\{\dots\}$; this does not matter since it commutes with multiplication by h. This ambiguity results from my attempts to avoid unnecessary parentheses and keep formulas readable.

2.4. COROLLARY. *Let A be a smooth projective real algebraic M-variety of dimension n, and let $\mathbb{C}\mathrm{A}$ and $\mathbb{R}\mathrm{A}$ be its sets of complex and real points respectively. Then for any pair of integers q, r such that $0 \leqslant q \leqslant r \leqslant n$ one has*

$$(\mathrm{in}_{2n})_! \left[V \cdot \mathrm{Sq}^{-1}\{h^{2q}(1+h)^{2n}\}\right]_{n+r} = (j_{2n})_! v_r(\mathbb{C}\mathrm{A}),$$

where $V = v(\tau\mathbb{R}\mathrm{A})v^{-1}(\tau\mathbb{R}\mathrm{A} \otimes \eta) \in H^*(\mathbb{R}\mathrm{A} \times \mathbb{R}\mathrm{p}^{2n})$.

REMARK. Corollary 2.4 expresses in terms of the characteristic classes of $\mathbb{C}\mathrm{A}$ the direct images in $\mathcal{S}^{2n}\mathbb{C}\mathrm{A}$ of the classes $\sum c^{ij}_{rq}(n)\omega_{ij}h^{n+r-i-j}$, where $c^{ij}_{rq}(n)$ are some constants, $c^{ij}_{rq}(n) \in \mathbb{Z}/2$, and $\omega_{ij} \in H^{i+j}(\mathbb{R}\mathrm{A})$ are certain characteristic classes of $\mathbb{R}\mathrm{A}$ (with $\omega_{00} = 1$). Sometimes the resulting system of equations can be resolved to give an expression for a class like $(\mathrm{in}_{2n})_! h^{2n-s}$, which, due to the fact that for an M-variety all the maps $(j_{kl})_!$ are monomorphisms, gives an expression for $(\mathrm{in}_s)_! 1$, i.e., the vanishing condition necessary for the existence of the canonical Stiefel orientations on $\mathbb{R}\mathrm{A}$ (see 1.2). To obtain more precise results, one needs deeper understanding of the coefficients $c^{ij}_{rq}(n)$, which depend on the expansions in h of V and $\mathrm{Sq}^{-1}\{h^{2q}(1+h)^{2n}\}$.

PROOF. We apply Theorem 2.3 to the complex conjugation involution on the manifold $X = \mathbb{C}\mathrm{A}$, $N = \dim_{\mathbb{R}} \mathbb{C}\mathrm{A} = 2n$, and use the fact that all the components of $F = \mathbb{R}\mathrm{A}$ have the same dimension n and that multiplication by $\sqrt{-1}$ gives an isomorphism $\nu\mathbb{R}\mathrm{A} = \tau\mathbb{R}\mathrm{A}$. □

3. Bredon operations and Hattori classes

3.1. Bredon operations (see Bredon [**B1**] for details). Recall that, given a paracompact topological space X, one can define the *external Steenrod operation* $P\colon H^*(X) \to H^*(\mathcal{S}^\infty_{\mathrm{tr}}(X \times X))$, where $X \times X$ is considered as a $\mathbb{Z}_2$-space via the transposing involution $\mathrm{tr}\colon (x,y) \mapsto (y,x)$. When combined with the diagonal map $\Delta\colon X \to X \times X$, $x \mapsto (x,x)$ (or, more precisely, with the map $\mathcal{S}^\infty_{\mathrm{id}} X \to \mathcal{S}^\infty_{\mathrm{tr}}(X \times X)$ induced by Δ), this operation gives the ordinary Steenrod squares (see Steenrod, Epstein [**SE**]): for a class $\alpha \in H^p(X)$ one has $\Delta^* P(\alpha) = \sum \mathrm{Sq}^i \alpha \cdot h^{p-i} \in H^*(\mathcal{S}^\infty_{\mathrm{id}} X) = H^*(X) \otimes \mathbb{Z}_2[h]$.

Suppose now that X is supplied with an involution c. Then one can combine P with the homomorphism induced by the twisted diagonal map $\Delta_c\colon X \to X \times X$, $x \mapsto (x, cx)$. The result, which is a homomorphism $(\Delta_c)^* P\colon H^*(X) \to H^*(\mathcal{S}^\infty_c X)$, is called the *Bredon operation* on X and is denoted by Br_c or just Br. Note that this operation is multiplicative, but, in general, not additive.

3.2. Hattori classes. Let X be an N-dimensional closed manifold. Then all the $\mathcal{S}^k X$, $k < \infty$, are also closed manifolds of dimensions $N + k$, and, hence, $H^{N+k}(\mathcal{S}^k X) = \mathbb{Z}_2$. Pick some $r \geqslant N/2$ and consider the composed map $\mathrm{Br}^{N-r}\colon H^r(X) \to H^{2r}(\mathcal{S}^\infty X) \to H^{2r}(\mathcal{S}^{2r-N} X) = \mathbb{Z}_2$. One can prove that Br^{N-r} is a homomorphism (see, e.g., [**L**]). Hence, due to Poincaré duality, there is a unique class $\theta_{N-r} \in H^{N-r}(X)$ such that $\mathrm{Br}^{N-r} \alpha = \theta_{N-r} \cdot \alpha$ for any $\alpha \in H^r(X)$.

3.3. DEFINITION. The above θ_{N-r} are called the *Hattori classes* of X. (We let $\theta_i = 0$ for $i > N/2$.) The sum $\theta = \sum \theta_i$ is called the *total Hattori class* of X.

REMARK. Note that, unlike ordinary characteristic classes, θ_0 may not be equal to $1 \in H^0(X)$.

3.4. THEOREM (Conner, Miller [**CM**]). *If c is a smooth involution on a smooth closed manifold X, then*

$$\theta = \mathrm{in}_!(v(\tau F)v^{-1}(\nu F)),$$

where τF and νF are the tangent bundle of F and normal bundle of F in X respectively, and v is the total Wu class of a vector bundle.

3.5. COROLLARY. $\theta_i = 0$ *if* $i < N - \dim F$ *or* $i > N/2$.

4. The Bredon operations in $\mathcal{S}^\infty X$

4.1. Since $\mathcal{S}^\infty X$ is also a $\mathbb{Z}_2$-space, one can consider $\mathcal{S}^\infty\mathcal{S}^\infty X$ and define Bredon operation $H^*(\mathcal{S}^\infty X) \to H^*(\mathcal{S}^\infty\mathcal{S}^\infty X)$. As in 2.1, $H^*(\mathcal{S}^\infty\mathcal{S}^\infty X)$ is an algebra over $H^*(\mathcal{S}^\infty\mathcal{S}^\infty \mathrm{pt}) = \mathbb{Z}_2[h] \otimes \mathbb{Z}_2[t]$.

4.2. THEOREM. *Let X be a paracompact space. Then:*

(1) *there is a canonical isomorphism $H^*(\mathcal{S}^\infty\mathcal{S}^\infty X) = H^*(\mathcal{S}^\infty X)\otimes\mathbb{Z}_2[t]$ compatible with the automorphism of the underlying ring $H^*(\mathbb{R}\mathrm{p}^\infty \times \mathbb{R}\mathrm{p}^\infty) = \mathbb{Z}_2[h,t]$ given by $h \mapsto h+t,\ t \mapsto t$;*

(2) *for a class $\alpha \in H^r(\mathcal{S}^\infty X)$ one has $\mathrm{Br}\,\alpha = \sum \mathrm{Sq}^i \alpha \cdot t^{r-i}$.*

PROOF. Consider $Y = S^\infty \times S^\infty$. The group $G = \mathbb{Z}_2 \times \mathbb{Z}_2$ acts on this space via the antipodal involutions a, a' on the factors. Let us re-decompose G into $\langle a\rangle \times \langle aa'\rangle$ and consider the G-space $Y' = S^\infty \times S^\infty$ on which a and aa' are the antipodal involutions on the first and second factor respectively. Since both Y and Y' are the total spaces of universal G-bundles, there are G-equivariant homotopy equivalences $\lambda\colon Y' \xrightarrow{\sim} Y$ and $\mathrm{id}_X\times\lambda\colon X \times Y' \xrightarrow{\sim} X \times Y$. Passing to the orbit spaces we obtain

$$\begin{aligned}\mathcal{S}^\infty\mathcal{S}^\infty X &= (X\times Y)/\langle ca, aa'\rangle \\ &\sim (X \times Y')/\langle ca, aa'\rangle = (X\times S^\infty)/\langle ca\rangle \times S^\infty/\langle aa'\rangle = \mathcal{S}^\infty X \times \mathbb{R}\mathrm{p}^\infty.\end{aligned}$$

This proves Statement (1).

To prove Statement (2), we apply similar arguments to classifying spaces of the group $H = \langle a_1, a_2, \mathrm{tr}\rangle/\{a_i^2 = \mathrm{tr}^2 = (a_1a_2)^2 = 1,\ \mathrm{tr}^{-1} a_1 \mathrm{tr} = a_2\}$. Let $Z = S^\infty \times S^\infty \times S^\infty$ be the H-space on which a_1, a_2 are the antipodal involutions on the first and second factor respectively and tr acts via $(\mathbf{t}_1, \mathbf{t}_2, \mathbf{u}) \mapsto (\mathbf{t}_2, \mathbf{t}_1, -\mathbf{u})$. Consider the homomorphism $\rho\colon G \to H$ given by $a \mapsto a_1a_2$, $a' \mapsto a_1a_2\,\mathrm{tr}$, and the ρ-equivariant maps

$$\begin{aligned}&\Delta\colon Y \longrightarrow Z, &&(\mathbf{t},\mathbf{u}) \mapsto (\mathbf{t}, -\mathbf{t}, \mathbf{u}), \quad\text{and}\\ &\Delta'\colon Y' \longrightarrow Z, &&(\mathbf{t},\mathbf{u}) \mapsto (\mathbf{t}, \mathbf{t}, \mathbf{u}).\end{aligned}$$

Since both $\Delta \circ \lambda$ and Δ' are classifying bundle maps corresponding to ρ, they are equivariantly homotopic. After we multiply everything by X (or $X \times X$), pass to the orbit spaces and quotient maps, and apply P, we obtain

$$\lambda^* \mathrm{Br} \overset{\mathrm{def}}{=\!=} \lambda^*(\Delta P) = \Delta' P \overset{\mathrm{def}}{=\!=} \widetilde{\mathrm{Sq}}, \quad \text{where } \widetilde{\mathrm{Sq}}\,\alpha = \sum \mathrm{Sq}^i \alpha \cdot t^{\dim\alpha - i}. \qquad \square$$

5. Hattori classes and Wu classes in $\mathcal{S}^\infty X$

5.1. THEOREM. *Let X be an N-dimensional closed manifold. Pick some integers $p, q, m \geqslant 0$ such that $2q \leqslant N+p$ and $m \geqslant N+2(p-q)$. Then*

$$(j_{pm})_!\theta_q(\mathcal{S}^pX) \overset{\infty}{=} \sum_{i=0}^{N+p-2q} \binom{p+i}{p}(j_{p+i,m})_!v_{q+i}(\mathcal{S}^{p+i}X),$$

where $\overset{\infty}{=}$ means that the two classes coincide modulo the annihilator

$$\operatorname{Ann}\operatorname{Im}[j_{m\infty}^*\colon H^{N+p-q}(\mathcal{S}^\infty X) \to H^{N+p-q}(\mathcal{S}^m X)].$$

5.2. COROLLARY. *Under the assumptions of Theorem 5.1, if $q \leqslant N/2$ then*

$$\sum_{p=0}^{N-2q} (j_{p,N-2q})_!\theta_{q+p}(\mathcal{S}^pX) \overset{\infty}{=} (j_{N-2q})_!v_q(X).$$

PROOF. Just sum up the statement of Theorem 5.1 and use the fact that $\sum_{p=0}^n \binom{n}{p} \equiv 0 \pmod 2$ if $n \neq 0$. $\square$

PROOF OF THEOREM 5.1. Let us denote by $[\![\alpha]\!]_k$ (resp., $[\![\alpha]\!]_{kl}$) the restriction to $H^{k+N}(\mathcal{S}^kX)$ (resp., to $H^{k+l+N}(\mathcal{S}^k\mathcal{S}^lX)$) of a given class $\alpha \in H^{k+N}(\mathcal{S}^\infty X)$ (resp., $\alpha \in H^{k+l+N}(\mathcal{S}^\infty\mathcal{S}^\infty X)$). Since $H^{k+N}(\mathcal{S}^kX) \cong H^{k+l+N}(\mathcal{S}^k\mathcal{S}^lX) \cong \mathbb{Z}_2$, we can regard $[\![\alpha]\!]_k$ and $[\![\alpha]\!]_{kl}$ as (mod 2)-residues and ignore their dimension; in particular, we can compare classes which belong to different groups.

We will prove that the values of the two classes in question on any class which comes from $H^{N+p-q}(\mathcal{S}^\infty X)$ coincide. (This means precisely that the two classes are congruent modulo the annihilator of $H^{N+p-q}(\mathcal{S}^\infty X)$; by the value of a cohomology class on another cohomology class of the complementary dimension we mean their cup-product evaluated on the fundamental class of the manifold.)

Put $r = N+p-q$, pick some $\alpha \in H^r(\mathcal{S}^\infty X)$, and consider $\operatorname{Br}\alpha$. By definition, $\alpha\cdot\theta_q(\mathcal{S}^pX) = [\![\operatorname{Br}\alpha]\!]_{r-q,p}$. According to Theorem 4.2, $\operatorname{Br}\alpha = \lambda^*\sum \operatorname{Sq}^i\alpha\cdot t^{r-i}$ for some isomorphism $\lambda^*\colon H^*(\mathcal{S}^\infty X)\otimes\mathbb{Z}_2[t] \to H^*(\mathcal{S}^\infty\mathcal{S}^\infty X)$ such that $\lambda^*(\beta t) = \lambda^*\beta\cdot t$ and $\lambda^*(\beta h) = \lambda^*\beta\cdot(h+t)$ for any $\beta \in H^*(\mathcal{S}^\infty X)$. If $i < q$, then t^{r-i} vanishes in $H^*(\mathcal{S}^{r-q}\mathcal{S}^pX)$, and the term $\operatorname{Sq}^i\alpha\cdot t^{r-i}$ does not contribute to $[\![\operatorname{Br}\alpha]\!]_{r-q,p}$. Because of dimensions, $j_{k\infty}^*\operatorname{Sq}^i\alpha = 0$ for $k < i+r-N = i+p-q$. Hence, $\operatorname{Sq}^i\alpha = \bar\alpha_i\cdot h^{i+p-q}$ for some $\bar\alpha_i \in H^N(\mathcal{S}^\infty X)$ such that $[\![\bar\alpha_i]\!]_0 = [\![\operatorname{Sq}^i\alpha]\!]_{i+p-q}$ (see Lemma 5.3 below), and we have

$$\lambda^*(\operatorname{Sq}^i\alpha\cdot t^{r-i}) = \lambda^*\bar\alpha_i\cdot(h+t)^{i+p-q}t^{r-i} = \lambda^*\bar\alpha_i\sum_{j=0}^{i+p-q}\binom{i+p-q}{j}h^j\,t^{r+p-q-j}.$$

The only term of this expansion which contributes to $[\![\lambda^*\operatorname{Sq}^i\alpha\cdot t^{r-i}]\!]_{r-q,p}$ is the one corresponding to $j=p$, the contribution being

$$\binom{i+p-q}{p}[\![\bar\alpha_i]\!]_0 = \binom{i+p-q}{p}[\![\operatorname{Sq}^i\alpha]\!]_{i+p-q} = \binom{i+p-q}{p}[\![\alpha\cdot v_i(\mathcal{S}^{i+p-q}X)]\!]_{i+p-q}.$$

(The latter identity is just the definition of $v_i(\mathcal{S}^{i+p-q}X)$.) To complete the proof it suffices now to sum up over $i = q, \dots, r$ and replace i with $i + q$. □

5.3. LEMMA. *The kernel* $\operatorname{Ker}[j^*_{k\infty}\colon H^*(\mathcal{S}^\infty X) \to H^*(\mathcal{S}^k X)]$ *coincides with the image* $\operatorname{Im}[\,\cdot h^{k+1}\colon H^*(\mathcal{S}^\infty X) \to H^*(\mathcal{S}^\infty X)]$.

PROOF. Consider the Serre spectral sequence E^{ij}_s of the fibration $\mathcal{S}^\infty X \to \mathbb{R}\mathrm{p}^\infty$. According to the construction, there is an isomorphism between the graded group associated with the filtration $\mathcal{F}^i = \operatorname{Ker}[j^*_{i\infty}\colon H^*(\mathcal{S}^\infty X) \to H^*(\mathcal{S}^i X)]$ on $H^*(\mathcal{S}^\infty X)$ and E^{**}_∞. Thus, it suffices to prove that all the maps $\cdot h^l\colon E^{ij}_s \to E^{i+l,j}_s$ are onto. This is obviously true in the term $E^{**}_2 = H^*(\mathbb{R}\mathrm{p}^\infty; H^*(X))$, and the general statement follows immediately from the following two observations, which can easily be proved by induction in s:

(1) $\cdot h^l\colon E^{ij}_s \to E^{i+l,j}_s$ is a monomorphism for $l \geqslant s-1$;
(2) if $\alpha = \beta \cdot h^l \in E^{i+l,j}_s$ and $d_s\alpha = 0$, then also $d_s\beta = 0$ (this follows from (1) and the fact that multiplication by h commutes with d_s). Thus, if α survives in E_{s+1}, then so does β, and still $\alpha = \beta \cdot h^l$. □

6. Proof of Theorem 2.3

6.1. We use the notation of 2.2: $[\alpha]_p$ is the p-dimensional component of a non-homogeneous class α, and $V = v(\tau F)v^{-1}(\nu F \otimes \eta) \in H^*(F \times \mathbb{R}\mathrm{p}^k)$, where η is the tautological linear bundle over $\mathbb{R}\mathrm{p}^k$, and τF and νF are the tangent bundle of F and normal bundle of F in X respectively.

6.2. PROPOSITION. *For any* $q \leqslant N/2$ *and* $p > N - q$ *one has*

$$\left[(\mathrm{in}_{N-2q})_! V \cdot \mathrm{Sq}^{-1}(1+h)^{N-2q+1}\right]_p = 0.$$

PROOF. The tangent bundle of $\mathcal{S}^k F$ and normal bundle of $\mathcal{S}^k F$ in $\mathcal{S}^k X$ are $\tau\mathcal{S}^k F = \tau F \times \tau\mathbb{R}\mathrm{p}^k$ and $\nu\mathcal{S}^k F = \nu F \otimes \eta$ respectively. Hence, Theorem 3.4 yields

$$\theta(\mathcal{S}^k X) = (\mathrm{in}_k)_! V \cdot \mathrm{Sq}^{-1}(1+h)^{k+1}, \tag{6.3}$$

and the statement follows from Corollary 3.5 applied to $\mathcal{S}^{N-2q}X$. □

6.4. PROPOSITION. *If* (X, c) *is an* M*-manifold, then for any* $q \leqslant N/2$ *one has*

$$\left[(\mathrm{in}_{N-2q})_! V \cdot \mathrm{Sq}^{-1}(1+h)^{N-2q}\right]_{N-q} = (j_{N-2q})_! v_q(X).$$

PROOF. If (X, c) is an M-manifold, from Smith theory (see, e.g., [**B2**]) it follows that all the maps j^*_{kl}, $0 \leqslant k \leqslant l \leqslant \infty$, are onto. Hence, the $\overset{\infty}{=}$-equality in Theorem 5.1 and Corollary 5.2 turns into identity, and, in particular, we have

$$\sum_{p=0}^{N-2q} (j_{p,N-2q})_! \theta_{q+p}(\mathcal{S}^p X) = (j_{N-2q})_! v_q(X).$$

Now we apply (6.3) and use the fact that

$$(6.5)\qquad (j_{kl})_!(\mathrm{in}_k)_!\alpha = (\mathrm{in}_l)_!\alpha \cdot h^{l-k} = (\mathrm{in}_l)_!\alpha \cdot \mathrm{Sq}^{-1}\{h(1+h)\}^{l-k}$$

for any $k \leqslant l$ and any class $\alpha \in H^*(\mathcal{S}^\infty F)$. Finally,

$$(j_{N-2q})_! v_q(X) = \left[(\mathrm{in}_{N-2q})_! V \cdot \sum_{p=0}^{N-2q} \mathrm{Sq}^{-1}\{(1+h)^{N-2q+1} h^{N-2q-p}\}\right]_{N-q}$$
$$= \left[(\mathrm{in}_{N-2q})_! V \cdot \mathrm{Sq}^{-1}(1+h)^{N-2q}\right]_{N-q}.$$

(We use the fact that $\sum_{i=0}^{N-2q} h^i = (1+h)^{-1}$ in $\mathcal{S}^{\leqslant N-2q}X$.) □

6.6. Proof of Theorem 2.3. Let us apply $(j_{N-2q,N})_!$ (see (6.5)) in order to put all the classes dealt with in Propositions 6.2 and 6.4 into $H^*(\mathcal{S}^N X)$. Denote

$$A_r^{(q)} = \left[(\mathrm{in}_N)_! V \cdot \mathrm{Sq}^{-1}\{h^{2q}(1+h)^N\}\right]_{N+r} \quad \text{and}$$
$$B_r^{(q)} = \left[(\mathrm{in}_N)_! V \cdot \mathrm{Sq}^{-1}\{h^{2q}(1+h)^{N+1}\}\right]_{N+r}.$$

These classes satisfy the obvious identity $A_r^{(q+1)} + B_r^{(q)} = A_r^{(q)} + hA_{r-1}^{(q)}$, which for $r > q$ turns into $A_r^{(q)} = A_r^{(q+1)} + hA_{r-1}^{(q)}$. (Proposition 6.2 states that $B_r^{(q)} = 0$ for $r > q$.)

We want to prove that $A_r^{(q)} = (j_N)_! v_r(X)$ for $N/2 \geqslant r \geqslant q \geqslant 0$. If $r = q$, this is the statement of Proposition 6.4. Hence, we can proceed by induction in $q = [N/2], [N/2]-1, \ldots$ and, for a fixed q, in $r = q, q+1, \ldots, [N/2]$. To complete the induction step it suffices to notice that, if $A_{r-1}^{(q)} = (j_N)_! v_{r-1}(X)$, then $hA_{r-1}^{(q)} = (j_N)_!\{hv_{r-1}(X)\} = 0$ (since $h = 0$ in $H^*(X)$), and the above identity turns into $A_r^{(q)} = A_r^{(q+1)} = (j_N)_! v_r(X)$. □

REMARK. Note that in fact, we proved that Theorem 2.3 is equivalent to Propositions 6.2 and 6.4.

REFERENCES

[B1] G. E. Bredon, *Cohomological aspects of transformation groups*, Proc. Conf. Transformation groups (New Orleans, La. 1967), Springer, 1968, pp. 245–280.

[B2] G. E. Bredon, *Introduction to compact transformation groups*, Academic Press, New York, London, 1972.

[CM] P. E. Conner, E. Y. Miller, *Equivariant self-intersection*, Preprint, 1979.

[D] A. Degtyarev, *Stiefel orientations on a real algebraic variety*, Real Algebraic Geometry (Rennes, 1991), Lect. Notes in Math., vol. 1524, Springer-Verlag, 1992, pp. 205–220.

[K] V. M. Kharlamov, *The maximal number of components of a surface of degree 4 in* $\mathbb{R}\mathrm{p}^3$, Functional Analysis and its Applications **6** (1972), no. 4, 101.

[L] T. Y. Lin, *Wu-like classes and generalized Peterson-Stein classes*, Fixed point theory (Sherbrooke, Que., 1980), Lect. Notes in Math., vol. 886, Springer, pp. 208–248.

[N] V. V. Nikulin, *Integral symmetric bilinear forms and some of their applications*, Izv. Akad. Nauk SSSR **43** (1979), no. 1, 117–177 (Russian); English transl. in Math. USSR-Izv. **14** (1980), 103–167.

[R] V. A. Rokhlin, *Complex topological characteristics of real algebraic surfaces*, Uspekhi Mat. Nauk **33** (1978), no. 4; English transl. in Russian Math. Surveys.

[SE] N. Steenrod, D. B. A. Epstein, *Cohomology operations*, Ann. Math. Studies, vol. 50, Princeton Univ. Press, 1962.

[V1] O. Viro, *Complex orientations of real algebraic surfaces*, Uspekhi Matematicheskikh Nauk **37** (1982), no. 4, 93.
[V2] O. Viro, *Complex orientations of real algebraic surfaces*, Geometry and Topology, I, Adv. Soviet Math., vol. 18, 1994.

UNIVERSITY OF CALIFORNIA AT RIVERSIDE
RIVERSIDE, CA 92521

E-mail address: degt@ucrmath.ucr.edu

Contemporary Mathematics
Volume **182**, 1995

EQUIVARIANT ALGEBRAIC REALIZATION OF SMOOTH MANIFOLDS AND VECTOR BUNDLES

KARL HEINZ DOVERMANN

ABSTRACT. Much of this article is a survey of results obtained as part of a program to develop real algebraic transformation groups. The emphasis is on the equivariant real algebraic realization of manifolds and vector bundles. We also prove our algebraic fixed point conjecture in some cases which we were not able to settle previously. Finally, we make some suggestions for future work.

0. Introduction

For some time the author has been interested in the problem of realizing smooth G manifolds and vector bundles algebraically, and much of this paper is a survey of the theory which we have developed so far. We also show some new results which confirm our algebraic fixed point conjecture in some cases which we were not able to settle previously. Finally, we make some suggestions for future work.

More specifically, in Section 1 we collect some important steps in the development of the non-equivariant theory. In Section 2 we describe how far we have been able to generalize these results in the presence of a group action. In Section 3 we discuss our fixed point conjecture, including its motivation as well as previous and new solutions. In Section 4 we focus on the algebraic realization of manifolds, exhibit some of the important tools, and derive a sample result. In Section 5 we state some structural results used in the verification of algebraic bundle realization, and derive sample results. In Section 6 we propose some ideas to improve on the results obtained so far, and suggest some directions in which to go in the future.

1. Some Non-equivariant Realization Results

The concept and definitions appealed to in this section are reasonably self-explanatory, and we will be precise once we enter the equivariant discussion.

As the starting point we quote a classical result of Seifert [Sf, p. 9].

Theorem 1.1. *If an orientable closed manifold $\mathfrak{M}^m$ is embedded twice differentiably in the Euclidean space $\mathbb{R}^n$, and if there are $(n-m)$ linearly independent continuous normals (according to §1 they always exist for $n-m=1$ and 2), then $\mathfrak{M}^m$ may be moved by an arbitrarily small isotopy into the real branch of an algebraic manifold*

1991 *Mathematics Subject Classification.* Primary 14P25, 57S15; Secondary 57S25.

In 1952 Nash generalized Seifert's result with a weaker conclusion. He proved that in general (without the assumption on the existence of the normals) M can be isotoped to a non-singular sheet of an algebraic variety in $\mathbb{R}^n$, and if $n \geq 2m+1$, then M can be isotoped to a non-singular component of an algebraic variety in $\mathbb{R}^n$. More importantly, the program set up in his paper, as well as the conjectures formulated in it, became one of the principal motivations to study real algebraic geometry. The other influencial paper of this time is by Whitney [Wh]. It develops foundations of algebraic geometry over the real numbers. Giving an answer to Nash's conjectures, Tognoli proved the following result [To].

Theorem 1.2. *A closed smooth submanifold of $\mathbb{R}^n$ may be deformed (in $\mathbb{R}^N$ for some $N > n$) by an arbitrarily small isotopy into a non-singular real algebraic variety*[1] .

The innovative step in this paper is that the problem can be reduced to a bordism problem; if the submanifold M of $\mathbb{R}^n$ is cobordant to a non-singular real algebraic variety, then it can be stably deformed into one by a small isotopy. Together with a result of Milnor [Mi1], this yielded the theorem.

The reduction of an algebraic realization problem to a bordism problem has become a common technique. E.g. King formulates this as follows [Kg].

General Principle. *If a topological situation is bordant to an algebraic situation, then it is isomorphic to an algebraic situation (if you have the right notion of bordism and do a lot of work).*

The first, unfortunately unpublished work on generalizing Nash's program to the equivariant setting, with, among others, the goal of getting an equivariant version of Tognoli's result, goes back to Palais [Pa].

Akbulut and King provided not only a more conceptual approach to the algebraic realization problem for manifolds, but their approach also opened the way to study the corresponding problem for manifolds which carry an additional structure [AK1], [AK2]. Among other results, and generalizing the result of Tognoli, they showed [AK1, Corollary 4.3]

Theorem 1.3. *Up the diffeomorphism, the set of non-singular real algebraic varieties is exactly the set of interiors of compact smooth manifolds with (possibly empty) boundaries.*

To simplify language, we say that a smooth manifold is *algebraically realized* if it is diffeomorphic to a non-singular real algebraic variety. The variety, or the variety together with the diffeomorphism, is then called an *algebraic realization* or *algebraic model* of the manifold.

Based on the approach of Akbulut and King, Benedetti and Tognoli showed [BT]

Theorem 1.4. *Every closed smooth manifold has an algebraic model X, such that every smooth vector bundle over X is strongly algebraic.*

More recently, Akbulut and King were able to get control over the need of stabilization in the theorems of Nash and Tognoli [AK3], [AK4].

---

[1] We use the expression 'real algebraic variety' for objects called 'real algebraic sets' by many of the authors we are referring to.

Theorem 1.5. *A closed smooth submanifold M of $\mathbb{R}^n$ can be deformed in $\mathbb{R}^n$ by an arbitrarily small isotopy into the non-singular points of a real algebraic variety, and as such it is the union of components of a real algebraic variety. In $\mathbb{R}^{n+1}$ it can be deformed by an arbitrarily small isotopy into a non-singular real algebraic variety.*

Once the question of algebraic realization is settled, the natural next question is whether, as in the transition from C^1 to C^∞ manifolds, there is a unique algebraic model on a closed smooth manifold. As shown by Bochnak and Kucharz, the answer is no [BK].

Theorem 1.6. *Every closed smooth manifold of dimensions ≥ 1 has an uncountable family of birationally inequivalent algebraic models.*

For more comprehensive discussions we like to refer the reader to a survey article of Ivanov [Iv] and books by Bochnak, Coste, and Roy [BCR] and Akbulut and King [AK5].

2. Equivariant Algebraic Realization

Our work on the equivariant realization of manifolds and vector bundles with group action has so far been motivated mostly by the results of Tognoli [To], Akbulut and King [AK1], Benedetti and Tognoli [BT], and Bochnak and Kucharz [BK] quoted above.

For completeness sake, we recall the definition of a real algebraic G variety[2]. Let G be a compact Lie group and Ω an orthogonal representation of G. Here we think of an orthogonal representation as an underlying Euclidean space $\mathbb{R}^n$ together with an action of G via orthogonal maps.

Definition 2.1. *A real algebraic G variety is the set of common zeros of polynomials $p_1, \ldots, p_m : \Omega \to \mathbb{R}$,*

$$V = \{x \in \Omega \mid p_1(x) = \cdots = p_m(x) = 0\},$$

which is invariant under the action of G. We also say that G acts real algebraically on V.

More exactly, such a variety should be called an affine variety, but as we are not considering any other varieties, such as projective ones, we omit this adjective throughout. An equivalent definition using algebraic groups and actions is given in Section 9 of [DM1]. The definition of *non-singularity* is as in the non-equivariant situation.

An example of importance to the discussion of strongly algebraic vector bundles is the Grassmannian $G_\Lambda(\Xi, k)$. Instead of defining it as a set of subspaces in a representation space we define it as a set of orthogonal projections onto subspaces. Let Λ stand for $\mathbb{R}$, $\mathbb{C}$, or $\mathbb{H}$. Let Ξ be a representation of G over Λ, in particular, its underlying space is Λ^n for some n. We assume that the action of G preserves the standard inner product (hermitian form) on Λ^n over Λ. Let $\mathrm{End}_\Lambda(\Xi)$ denote

[2]For readers who like to see more background material on the subject of real algebraic transformation groups we suggest an older survey article [DS2].

the set of endomorphisms of Ξ over Λ. It is a representation of G with the action given by

$$G \times \mathrm{End}_\Lambda(\Xi) \to \mathrm{End}_\Lambda(\Xi) \quad \text{with} \quad (g, L) \mapsto gLg^{-1}.$$

Let k be a natural number. We set

$$G_\Lambda(\Xi, k) = \{L \in \mathrm{End}_\Lambda(\Xi) \mid L^2 = L,\ L^* = L,\ \mathrm{trace}\, L = k\}$$
$$E_\Lambda(\Xi, k) = \{(L, u) \in \mathrm{End}_\Lambda(\Xi) \times \Xi \mid L \in G_\Lambda(\Xi, k),\ Lu = u\}.$$

Here L^* denotes the adjoint of L. If one chooses an orthonormal (resp. unitary or symplectic) basis of Ξ, then $\mathrm{End}_\Lambda(\Xi)$ is canonically identified with the set of $n \times n$ matrices Λ^{n^2}, and L^* is obtained by transposing L and conjugating its entries. This description specifies $G_\Lambda(\Xi, k)$ and $E_\Lambda(\Xi, k)$ as real algebraic G varieties. Define $p : E_\Lambda(\Xi, k) \to G_\Lambda(\Xi, k)$ as projection on the first factor. This defines a G vector bundle, which is called the *universal bundle* over $G_\Lambda(\Xi, k)$ and denoted by $\gamma_\Lambda(\Xi, k)$.

In analogy with the non-equivariant situation, let us say that a smooth G manifold is algebraically realized if it is equivariantly diffeomorphic to a non-singular real algebraic G variety. As a generalization of Tognoli's theorem one is tempted to make the

Algebraic Realization Conjecture. *Let G be a compact Lie group. Every closed smooth G manifold is algebraically realized.*[3]

Now, in analogy the the situation without group action, a smooth G manifold is said *algebraically realized* if it is equivariantly diffeomorphic to a non-singular real algebraic G variety.

Using various techniques we have been able to verify this conjecture in a number of cases. We will discuss some structural elements of the proof later.

Theorem A. *Let G be a compact Lie group. A closed smooth G manifold is algebraically realized if one of the following assumptions holds.*

(1) *G is the product of a group of odd order and a 2-torus.*
(2) *G is finite abelian with cyclic 2-Sylow subgroup or G is S^1.*
(3) *The action of G on the manifold is semifree.*
(4) *The manifold is of dimension two.*

The proofs of the different items are scattered over several references. Many cases of (1) are proved in [DM1], and in the stated generality the result appears in [DMS1]. Item (2) is one of the major results in [DM2]. (3) also appears in [DM1]. We like to remind the reader that the action on M is called semifree if for all $x \in M$ the isotropy group $G_x = \{g \in G \mid gx = x\}$ is either trivial or all of G. This includes the case of free actions where $G_x = 1$ for all $x \in M$. For orientable actions (4) is shown in [DKS], in general it is shown in [KM].

[3]Throughout this paper we follow the convention that a manifold has a dimension, i.e., all of its components have the same dimension. In part, this is motivated by the same convention for non-singular varieties. This conjecture and the results of this paper can be applied to manifolds with components of different dimension by applying them separately to the components of a fixed dimension, and then taking a disjoint union.

Unlike in the non-equivariant theory, in the equivariant setting the algebraic realization of bundles enters even into the algebraic realization of manifolds. Before we can discuss this topic, we need to define morphisms in the algebraic category. Let $V \subset \mathbb{R}^n$ and $W \subset \mathbb{R}^m$ be real algebraic varieties. A map $f : V \to W$ is said to be *regular* if it extends to a map $F : \mathbb{R}^n \to \mathbb{R}^m$ such that each component of F is a polynomial. We say that f is *entire rational* if there are regular maps $p : \mathbb{R}^n \to \mathbb{R}^m$ and $q : \mathbb{R}^n \to \mathbb{R}$ such that $f = p/q$ on V and q does not vanish on V. If V and W are G varieties and f is in addition equivariant, then we call f an *equivariant regular map*, resp., an *equivariant entire rational map.* The following is easy to show, see [DMS1, Section 2]. Let $V \subset \Omega$ and $W \subset \Xi$ be real algebraic G varieties and $f : V \to W$ equivariant regular map. Then the regular extension $F : \Omega \to \Xi$ of f may be assumed to be equivariant. If $f : V \to W$ is an equivariant entire rational map, then there exist equivariant polynomials $P : \Xi \to \Omega$ and $Q : \Xi \to \mathbb{R}$ such that $f = P/Q$, restricted to V, and Q does not vanish on Ξ.

In our context it is useful to consider strongly algebraic G vector bundles as well as algebraically realized G vector bundles, i.e., bundles on which we can impose a strongly algebraic structure.

Definition 2.1. *A strongly algebraic G vector bundle over Λ is a pair (X, μ) where X is a real algebraic G variety and $\mu : X \to G_\Lambda(\Xi, k)$ is an equivariant entire rational map.*

The advantage of our definition of a strongly algebraic bundle is that one does not have to define algebraic bundles first, as done in [BT] and other papers dealing with this concept. As above, Λ stands for $\mathbb{R}$, $\mathbb{C}$, or $\mathbb{H}$. Depending on the choice of Λ, we are considering real, complex, or quaternion bundles.

Definition 2.2. *Let M be a smooth G manifold, $\mathfrak{F}$ a collection of G vector bundles over M, X a non-singular real algebraic variety, and $\varphi : X \to M$ an equivariant diffeomorphism. We say that (X, φ) provides an algebraic realization of $\mathfrak{F}$ if $\chi_\xi \circ \varphi$ is equivariantly homotopic to an entire rational map for each $\xi \in \mathfrak{F}$, where χ_ξ is a classifying map for ξ. Given M and $\mathfrak{F}$ as before, we say that $\mathfrak{F}$ is algebraically realized if there exists a pair (X, φ) as above which provides an algebraic realization of $\mathfrak{F}$.*

In [DMS1] we showed the following result

Theorem B. *The set of all real G vector bundles over a closed smooth G manifold is algebraically realized if one of the following assumptions holds:*

(1) *G is the product of a group of odd order and a 2-torus.*
(2) *The action of G on the manifold is semifree.*

As in the non-equivariant setting, we can now ask how many different algebraic models a closed smooth G manifold has. Assuming that there is at least one such model, we obtain the following result. It is proved in [DMS2] and generalizes the result of Bocknak and Kucharz [BK] mentioned above.

Theorem C. *Let G be a finite group and X a compact non-singular real algebraic G variety of dimension at least 1. Then there exists an uncountable family $\{X_\alpha\}_{\alpha\in\Lambda}$*

of non-singular real algebraic G varieties such that

(1) *considered as smooth G manifolds, X_α is equivariantly diffeomorphic to X and*
(2) *X_α is not birationally equivalent to X_β for $\alpha \neq \beta$.*

In constrast, we have shown in [DM2]:

Theorem D. *Let G be a compact Lie group and H a closed subgroup. Then G/H has the structure of a non-singular real algebraic G variety. This structure is unique. If H′ is another closed subgroup of G and $H' \supset H$, then the induced map $G/H \to G/H'$ is entire rational.*

It is important to point out that the uniqueness part of the statement is a consequence of the requirement that G acts real algebraically on G/H. The result was pointed out to us by F. Knop, who calls it a corollary of the Peter-Weyl theorem. G. Schwarz helped us with the details of the proof. In our effort to understand the background better, we discovered a proof in the unpublished manuscript of Palais mentioned earlier [Pa, p. 154]. As a variety, G/H is realized as an orbit of a point v with isotropy group H in a representation of G, this one deduces from results in [Sw1]. Its ring of polynomial functions, which determines the algebraic structure on G/H, is $\mathbb{R}[G]^H$. A fairly 'real' proof is obtained from the orthogonality theorem and the Stone Weierstrass theorem (e.g., see [Ho]). A fairly 'complex' proof may be obtained using that, in the complex theory, a quotient of a linear algebraic group by a closed subgroup has a unique algebraic structure [Bo], [Hu]. Applying Matsushima's Theorem [BH] and Birkes' theorem [Bi] to the complexifications of G and H, the quotient will be affine, and its real points are G/H. More details on the proof can be found in [DM2], and a revised version of this reference should treat the topic in a broader context.

3. The Fixed Point Conjecture

The fixed point conjecture is the problem which motivated our work in real algebraic transformation groups. Its origins are in complex algebraic transformation groups. We will state the fixed point conjecture and demonstrate in which way the real algebraic theory is quite different from the smooth theory for non-compact settings. Then we state the currently available solutions of the conjecture and give a detailed proof for the case which is new. Finally, we discuss the conjectures in complex algebraic transformation groups which led to ours and what happened to them.

Fixed Point Conjecture (see [DMP, p. 50]). *A compact Lie group acts real algebraically and without a fixed point on a variety diffeomorphic to Euclidean space if and only if it acts smoothly and without a fixed point on a disk.*

The word *disk* is used for the points of norm ≤ 1 in some Euclidean space. The necessity part ($\Longrightarrow$) has been shown by Petrie and Randall [PR1], [P1], see also [PR2]. The basic idea is as follows. Given a real algebraic variety (at least it needs to be smooth) with an algebraic action of a compact Lie group G, there exists an equivariant polynomial function $f : X \to \mathbb{R}$ which looks very much like a Morse function. Because it is a polynomial it has only finitely many critical values. For a sufficiently large N, $X_N := f^{-1}(-\infty, N]$ is a compact smooth G manifold, and $X = X_N \cup_{\partial X_N} (\partial X_N \times [0,\infty))$. Assuming X is contractible it follows that X_N is diffeomorphic to a disk. Given that the action on X has no fixed point it follows that the action on X_N has no fixed point. Petrie and Randall's work is in the construction of f and the verification that f has basically all the properties of a Morse function.

One may want to compare the argument in the previous paragraph with the one used by Akbulut and King in their proof of the result which we stated as Theorem 1.3 in the introduction [AK1, p. 446]. In both cases one compactifies by attaching a boundary. At this point, quite contrary to the algebraic realization conjecture, the smooth and real algebraic theory behave rather differently. Being algebraic imposes a severe restriction not encountered in the smooth category. This can be observed in trivial cases; $\mathbb{R} \setminus \mathbb{Z}$ is a smooth manifold, but not algebraically realized. The fixed point conjecture expresses this difference in an equivariant setting. To see the difference we need to know what happens in the smooth situation. Oliver showed (see [O])

Theorem 3.1. *A finite group G acts smoothly and without a fixed point on a disk if and only if there does not exist a normal series $P \lhd H \lhd G$ such that P and G/H are of prime power order and H/P is cyclic. (E.g., a finite abelian group acts without a fixed point on a disk if and only if it has at least three non-cyclic Sylow subgroups.)*

On the other hand, for cyclic groups not of prime power order Conner and Floyd [CF] constructed fixed point free actions on $\mathbb{R}^n$. Additional work by Conner and Montgomery [CM], Hsiang and Hsiang [HH], and Edmonds and Lee [EL] (apply the Mostow-Palais embedding theorem to give smooth action where the authors don't) has the following cummulative result (the trivial group is considered to be a group of prime power order).

Theorem 3.2. *A compact Lie group G with connected component G_0 has a fixed point free action on some $\mathbb{R}^n$ if and only if G/G_0 is not of prime power order or G_0 is not abelian.*

Let us call a finite group G an *Oliver group* (this expression was coined by Laitinen and Morimoto [LM]) if it has a smooth fixed point free action on a disk.

Theorem 3.3. *The alternating group A_5 and odd order Oliver groups act real algebraically and without a fixed point on varieties diffeomorphic to a Euclidean space.*

For A_5 this has been shown in [DMP], see also [DM3] and for odd order abelian Oliver groups in [DKS] (in combination with results from [DS1]), see also [DM1]. We will give the proof for odd order Oliver groups in general below.

Let us draw some conclusions for the fixed point conjecture. We are only concerned with the sufficiency because the necessity was shown by Petrie and Randall. It follows from Oliver's theorem (stated above) that the symmetric group S_k and the alternating group A_k act smoothly on a disk without a fixed point if and only if $k \geq 5$. Earlier, Floyd and Richardson (see [FR] or [Br, p. 58]) had shown that A_5 acts smoothly on a disk without a fixed point, and it follows from an elementary construction that A_k and S_k do as well if $k \geq 5$. Observe also

Proposition 3.4 (see [DMP]). *Let $\mathcal{G}$ denote the set of all compact Lie groups which act real algebraically, effectively, and without fixed point on a variety diffeomorphic to Euclidean space, and let G be a compact Lie group.*

(1) *If $H \subset G$ is a subgroup of finite index and $H \in \mathcal{G}$, then $G \in \mathcal{G}$.*
(2) *If G surjects onto K and $K \in \mathcal{G}$, then $G \in \mathcal{G}$.*

Based on this discussion and Theorem 3.3 we conclude:

Corollary E. *The fixed point conjecture holds for symmetric groups, alternating groups, and odd order groups.*

Taken together, Proposition 3.4 and Corollary E provide a large class of groups for which the fixed point conjecture holds.

Let us state two results which are the major steps in the proof of Theorem 3.3.

Theorem 3.5. *A finite group G has a smooth action on a sphere with exactly one fixed point if (and only if) G is an Oliver group.*

This is a theorem of Laitinen and Morimoto [LM]. The proof depends heavily on so far unavailable work of Bak and Morimoto [BM]. For this reason we decided to write out a proof in the special cases needed in the proof to Theorem 3.3, see the appendix to this paper, where one can rely on published results instead of the work of Bak and Morimoto [BM].

Lemma 3.6 (see [DMP, p. 50] and compare [M2, p. 105]). *Let V be a non-singular real algebraic G variety and $x \in V$ a fixed point. Then $V \setminus x$ is equivariantly diffeomorphic to a non-singular real algebraic G variety. If V is a homotopy sphere and the action has exactly one fixed point, then $V \setminus x$ is diffeomorphic to a Euclidean space on which the action has no fixed point.*

Proof of Theorem 3.3. Let G be an odd order Oliver group. It acts smoothly and with exactly one fixed point on a homotopy sphere, see Theorem 3.5. Theorem A provides us with a real algebraic structure for this action. Then we apply Lemma 3.6 to obtain the real algebraic action on a variety diffeomorphic to a Euclidean space with empty fixed point set. □

Next we explain the (complex) algebraic background of the fixed point conjecture. We provide only some of the relevant definitions, but hope that the reader gets at least a fairly accurate understanding of the problems. For further reading we suggest the given references. We suppose that G is a reductive group [Kr1], such as a finite group, the non-zero complex numbers $\mathbb{C}^*$ with multiplication as group operation, or the group $O(2, \mathbb{C})$ consisting of orthogonal matrices with entries in $\mathbb{C}$. The following two conjectures were of much interest in algebraic transformation groups.

Equivariant Serre Conjecture[4]**.** *Every complex algebraic G vector bundle over a representation space is isomorphic to a product bundle.*

[4] It appears that Bass is the first person to attach this name to the conjecture, and in analogy to the situation without group action, the name seems to be justified. Still, Serre never made this conjecture.

Note that this conjecture holds for topological vector bundles because a representation space contracts to a point. An algebraic bundle is defined just like a topological one, only that one requires all data (such as the transition functions) to be algebraic, and local triviality is assumed over Zariski open sets.

Linearity Conjecture [Ka]. *Every algebraic action of a reductive group on $\mathbb{C}^n$ is conjugate to a linear action.*

Suppose $\mu, \eta : G \times \mathbb{C}^n \to \mathbb{C}^n$ are algebraic actions. They are said to be *conjugate* to each other (in the complex algebraic category) if there exists an algebraic automorphism $\psi : \mathbb{C}^n \to \mathbb{C}^n$ such that the following diagram commutes

$$\begin{array}{ccc} G \times \mathbb{C}^n & \xrightarrow{\mu} & \mathbb{C}^n \\ {\scriptstyle Id\times\psi}\big\downarrow & & \big\downarrow{\scriptstyle \psi} \\ G \times \mathbb{C}^n & \xrightarrow{\eta} & \mathbb{C}^n \end{array}$$

In the topological category such a conjecture is certainly not true. As we mentioned earlier, there are even cyclic actions on $\mathbb{R}^n$ without a fixed point, while every linear action leaves at least the origin fixed. A negative answer to the Fixed Point Conjecture in the corresponding complex algebraic setting, and on $\mathbb{C}^n$, would have been a striking counter example to the Linearity Conjecture.

The situation is as follows. Let E be the total space of an algebraic vector bundle ξ over a representation space of a reductive group G. According to the solution of the Serre Conjecture by Quillen and Suslin, E is (algebraically isomorphic to) $\mathbb{C}^n$. The vector bundle structure defines an additional action of $\mathbb{C}^*$ on E, so that $G \times \mathbb{C}^*$ acts on the total space of the bundle. If ξ is nontrivial as a G vector bundle, then Kraft showed that the action of $G \times \mathbb{C}^*$ on E is not conjugate to a linear action [Kr2]. In a fundamental paper Schwarz produced counter examples to the Equivariant Serre Conjecture for groups such as $O(2, \mathbb{C})$, hence counter examples to the Linearity Conjecture for groups such as $O(2, \mathbb{C}) \times \mathbb{C}^*$ [Sw2] (for a detailled account with complete proofs, see [KS]). Continuous families of such examples for finite groups were given by Masuda and Petrie [MP], the complete paper is still in preprint form. For related results see also [Kn] and [MMP]. There is also a survey article by Masuda on real algebraic transformation groups with focus on the problem of linearizing affine algebraic actions [Ma].

Miki considered the corresponding problems in the real setting and constructed uncountably many real algebraic actions of $O(2, \mathbb{R})$ on $\mathbb{R}^4$ [Mk]. These are regular actions, and they are inequivalent in this category. As rational representations they are equivalent.

4. Ingredients for Manifold Realization

First of all, we need an equivariant theorem which expresses the general principle from Section 1. It allows to reduce realization problems to bordism problems.

Let $f_i : M_i \to Y$ be equivariant maps whose domain is a closed smooth G manifold, $i = 1, 2$. They are said to be *equivariantly cobordant* if there exists a compact smooth G manifold W and an equivariant map $F : W \to Y$, such that the boundary of W is the disjoint union of M_1 and M_2 and F restricts to f_i on

M_i. Suppose that Y is a real algebraic G variety. We say that $f : M \to Y$ is *algebraically realized* if there exists a non-singular real algebraic G variety X, an equivariant entire rational map $\mu : X \to Y$, and an equivariant diffeomorphism $\phi : M \to X$, such that $\mu \circ \phi$ is equivariantly homotopic to f. We say that (M, f) can be *approximated algebraically* if there are algebraic realizations (X, μ, ϕ) such that $\mu \circ \phi$ is arbitrarily close to f in the C^1 topology [Hi]. An equivariant bordism class is said to be *algebraically realized* if it has a representative $\mu : X \to Y$ where X is a non-singular real algebraic G variety and μ is an equivariant entire rational map. In [DMS1, Theorem C and its addendum in Section 7] we showed the following generalization of results from [AK1].

Theorem 4.1. *Let G be a compact Lie group. An equivariant map from a closed smooth G manifold to a non-singular real algebraic G variety can be approximated algebraically if and only if its equivariant bordism class is algebraically realized.*

Professor Richard Palais told us that in the 70's, he and Henry King thought about a theorem similar to our Theorem 4.1 in the special case where Y is a point, but they never wrote out a detailed argument.

Example. Just using this theorem we can handle one easy situation. Suppose the group G has a central element τ of order 2, and M is a smooth G manifold such that the action of τ is free. We show that M is algebraically realized. To see this, observe that the orbit space M/τ (identify x with τx) is a smooth manifold (use that τ acts freely), G acts smoothly on M/τ (use that τ is central), and the natural projection $\pi : M \to M/\tau$ is equivariant. The mapping cylinder of π is a smooth G manifold (use that τ is of order 2), and its boundary is M. Hence M bounds equivariantly (it is cobordant to the empty set, which is algebraic), and Theorem 4.1 (applied with Y a point) implies that M is algebraically realized.

Let us focus for the moment on equivariant manifold realization. The equivariant bordism classes of closed smooth G manifolds with disjoint union and cartesian product of representatives as addition and multiplication form a graded ring $\mathfrak{N}_*^G$ and a module over $\mathfrak{N}_*$ (the bordism ring of the trivial group). It is elementary to observe

Proposition 4.2. *Let G be a compact Lie group and $\mathfrak{S}$ a set of $\mathfrak{N}_*$ module generators of $\mathfrak{N}_*^G$. If each element of $\mathfrak{S}$ is algebraically realized, then every closed smooth G manifold is algebraically realized.*

This proposition is useful as typically the results of the available calculations of equivariant bordism groups are given in these terms. Typically, the generators are also total spaces of projective bundles of vector bundles, and the vector bundles are built up from simple ones with the help of operations like Whitney sum and tensor product. This is the motivation for our next two propositions from [DM1] and [DMS1]. In addition, induction techniques enter, and these we studied in [DM1]. We state the result as another proposition.

Proposition 4.3. *Let $(B, \alpha : B \to G_{\mathbb{R}}(\Xi, k))$ be a strongly algebraic G vector bundle. The total space*

$$\mathbb{R}P(E) = \{(x, T) \in B \times G_{\mathbb{R}}(\Xi, 1) \mid (Id - \alpha(x))T = 0\}$$

of the associated projective bundle is a real algebraic G variety and the natural projection map $p : \mathbb{R}P(E) \to B$ is an equivariant entire rational map. If B is non-singular, then $\mathbb{R}P(E)$ is non-singular.

Proposition 4.4. *Let X be a real algebraic G variety and $\mu : X \to G_\Lambda(\Xi, k)$ and $\mu' : X \to G_\Lambda(\Xi', k')$ be two strongly algebraic G vector bundles over Λ. Then*

(1) *The Whitney sum $\mu \oplus \mu' : X \to G_\Lambda(\Xi \oplus \Xi', k + k')$ is a strongly algebraic G vector bundle over Λ.*
(2) *The tensor product $\mu \otimes \mu' : X \to G_{\Lambda'}(\Xi \otimes_\Lambda \Xi', \epsilon_\Lambda \cdot k \cdot k')$ is a strongly algebraic G vector bundle over Λ'. Here $\Lambda' = \mathbb{R}$ if $\Lambda = \mathbb{R}$ or $\mathbb{H}$, and $\Lambda' = \mathbb{C}$ if $\Lambda = \mathbb{C}$, and $\epsilon_\Lambda = 1$ if $\Lambda = \mathbb{R}$ or $\mathbb{C}$, and $\epsilon_\Lambda = 4$ if $\Lambda = \mathbb{H}$.*
(3) *The dual μ^* of μ is a strongly algebraic G vector bundle over $\mathfrak{k} = \mathbb{R}$ if $\Lambda = \mathbb{R}$ or $\mathbb{H}$, and over $\mathfrak{k} = \mathbb{C}$ if $\Lambda = \mathbb{C}$. Furthermore,* $\mathrm{Hom}(\mu, \mu')$ *(which is isomorphic to $\mu^* \otimes \mu'$) is strongly algebraic over $\mathfrak{k}$, where $\mathfrak{k}$ is as in the previous sentence.*
(4) *If μ' is a strongly algebraic subbundle of μ, then the orthogonal complement of μ' in μ is strongly algebraic.*
(5) *Every strongly algebraic G vector bundle over Λ may be viewed as a strongly algebraic G vector bundle over $\mathbb{R}$.*
(6) *If Γ is a k-dimensional representation of G over Λ, then the product bundle over X with fibre Γ, represented by the constant map to any point in the component of $G_\Lambda(\Xi, k)^G$ associated to the representation Γ, is a strongly algebraic G vector bundle over Λ.*
(7) *In the assumption and conclusion for (1)–(6) we may replace strongly algebraic G vector bundles by algebraically realized bundles. But in (1)–(3) we need to assume that $\mu = (E, p, B)$ and $\mu' = (E', p', B)$ are algebraically realized over the same non-singular variety X.*

Item (4) may need some explanation. Let $\mu : X \to G_\Lambda(\Xi, k)$. We say that $\mu' : X \to G_\Lambda(\Xi, k')$ is a subbundle if $k' \le k$ and $\mu(x) \circ \mu'(x) = \mu'(x)$. Both, μ and μ', are assumed to be equivariant entire rational maps. The orthogonal complement $\mu^\perp : X \to G_\Lambda(\Xi, k - k')$ of μ' in μ is then given by $\mu^\perp(x) = (Id - \mu'(x)) \circ \mu(x)$. This is the analog of the 'orthogonal complement' construction for bundles in terms of their classifying maps.

Let H be a subgroup of a group G, and let X be an H space. The induced G space $\mathrm{Ind}_H^G X$ is defined as the orbit space $G \times_H X$. More explicitly, elements in $G \times_H X$ are equivalence classes of elements in $G \times X$, where (g, x) is equivalent to $(gh, h^{-1}x)$ for all $h \in H$. If X is a (closed) smooth H manifold and H is a closed subgroup of a compact Lie group G, then $\mathrm{Ind}_H^G X$ is a (closed) smooth G manifold. This procedure is functorial. Given an H equivariant map $f : X \to Y$ of H spaces, the map $Id \times f : G \times X \to G \times Y$ induces a G map $\mathrm{Ind}_H^G f : \mathrm{Ind}_H^G X \to \mathrm{Ind}_H^G Y$ of orbit spaces. If X and Y are smooth H manifolds and f is smooth, then so is $\mathrm{Ind}_H^G f$. In [DM1] we proved a similar result for real algebraic varieties.

Proposition 4.5. *Suppose G is a compact Lie group, and H is a closed subgroup of finite index. If X is a (non-singular) real algebraic H variety, then there is a natural procedure to define a (non-singular) real algebraic G variety structure on $\mathrm{Ind}_H^G X$. Given two real algebraic H varieties X and Y and an H equivariant regular (resp.*

entire rational) map $f : X \to Y$, *there is a naturally defined* G *equivariant regular (resp. entire rational) map* $\mathrm{Ind}_H^G f : \mathrm{Ind}_H^G X \to \mathrm{Ind}_H^G Y$.

These tools suffice to show Theorem A for odd order abelian groups. One can find an explicit description of the generators of $\mathfrak{N}_*^G$ (G of odd order) as a module over $\mathfrak{N}_*$ in [Co] or [Ko] (it can also be extracted from [St1]). They are disjoint unions of manifolds of the form

$$\overline{R_{H,V,i}} = G \times_H \mathbb{R}P((\eta_{i-1} \otimes_{\mathbb{C}} \underline{V}) \oplus \underline{\mathbb{R}}).$$

Here η_{i-1} is the canonical line bundle over $\mathbb{C}P^{i-1}$ with trivial action of H and H is a subgroup of G. This bundle is tensored with the product bundle $\underline{V}$ over $\mathbb{C}P^{i-1}$ with fibre V. We give V any of its complex structures. Then we add the product bundle $\underline{\mathbb{R}}$ with fibre $\mathbb{R}$ and take the total space of the associated projective bundle. Finally, we take the balanced product with G over H. The result is $\overline{R_{H,V,i}}$. Using 4.3–4.5 one sees that each $\overline{R_{H,V,i}}$ is algebraically realized, so 4.2 implies that for odd order abelian groups every closed smooth G manifold is algebraically realized.

5. Ingredients of Bundle Realization

Given a closed smooth G manifold, the set of all real G vector bundles over M is algebraically realized it there exists an algebraic realization (X, φ) of M such that for every real representation Ξ of G and every natural number k, every equivariant map $\chi : X \to G_{\mathbb{R}}(\Xi, k)$ is equivariantly homotopic to an entire rational map. The next two propositions reduce the problem, first to a K-theoretical problem and then to a problem for a finite number of bundles, see [DMS1, Proposition 2.13 and Lemma 2.14]. The first one generalizes a result of Benedetti and Tognoli [BT], the second one uses that equivariant K theory is finitely generated [Sg].

Lemma 5.1. *Let* G *be a compact Lie group,* X *a compact non-singular real algebraic* G *variety, and* ξ, ξ_0, *and* ξ_1 *a real* G *vector bundles over* X. *Suppose that the classifying maps of* ξ *and* ξ_0 *are equivariantly homotopic to entire rational maps, and* $\xi_0 \oplus \xi_1 \cong \xi$, *then the classifying map of* ξ_1 *is equivariantly homotopic to an entire rational map.*

Proposition 5.2. *Let* G *be a compact Lie group and* M *a closed smooth* G *manifold. Suppose every finite collection of* G *vector bundles over* M *can be algebraically realized, then the set of all* G *vector bundles over* M *can be algebraically realized.*

For a closed smooth G manifold M we would now take a finite set $\mathfrak{F} = \{\xi_1, \dots, \xi_k\}$ of G vector bundles and their classifying maps $\chi_j : M \to G(\xi_j)$, $1 \le j \le k$, where $G(\xi_j)$ stands for an appropriate Grassmann variety, and try to algebraically realize the product $\chi_{\mathfrak{F}}$ of the χ_j,

$$\chi_{\mathfrak{F}} : M \to G(\mathfrak{F}) := G(\xi_1) \times \cdots \times G(\xi_k).$$

Theorem 4.1 reduces this problem to an equivariant bordism problem, and this may be formulated more generally as

Algebraic Bordism Problem for Equivariant Maps. *Let Y be a real algebraic G variety. Is every homogeneous class $\alpha \in \mathfrak{N}_*^G(Y)$ algebraically realized?*

Even in the non-equivariant setting the answer to this problem depends in a subtle way on the algebraic structure of Y. Akbulut and King introduce the idea of a variety having *totally algebraic homology* [AK1]. It is a consequence of their paper that for a non-singular variety Y every homogeneous class $\alpha \in \mathfrak{N}_*^G(Y)$ is algebraically realized if and only if Y has totally algebraic homology (see [AK1, Lemma 2.5]). Only few varieties, such as Grassmannians with their natural algebraic structure, are known to have totally algebraic homology. Benedetti and Dedò showed that there are closed smooth manifolds (in all dimensions ≥ 11) such that none of their algebraic realizations has totally algebraic homology [BD, Theorem 2]. On the other hand, Akbulut and King have shown that every closed smooth manifold is homeomorphic to a real algebraic variety with totally algebraic homology [AK6]. Necessarily, this variety must some times have singularities. In the absence of a Künneth formula in equivariant bordism, which is used to reduce the bordism problem to a homology problem, it seems most appropriate to say:

Definition. *A real algebraic G variety Y has totally algebraic bordism if every homogeneous class in $\mathfrak{N}_*^G(Y)$ is algebraically realized.*

An analysis of the calculation of the equivariant bordism group for odd order groups, together with the analysis of the fixed point structure of $G_{\mathbb{R}}(\Xi, k)$ allows us to deduce the result needed to prove Theorem B in case G is of odd order (see [DMS1, Proposition 3.3]).

Proposition 5.3. *Let G be a group of odd order, Ξ_i orthogonal representations of G and k_i natural numbers, $i = 1, \ldots, m$. The G variety $\prod_{i=1}^m G_{\mathbb{R}}(\Xi_i, k_i)$ has totally algebraic bordism, i. e., every homogeneous class in $\mathfrak{N}_*^G(\prod_{i=1}^m G_{\mathbb{R}}(\Xi_i, k_i))$ is algebraically realized.*

To obtain Theorem B for odd order groups times 2 tori, one uses the the the above and the following localization result. Its proof generalizes techniques used by Stong [St2].

Proposition 5.4. *Let G be a compact Lie group, $\tau \in G$ a central element of order two in G and M a closed smooth G manifold. The set of all real G vector bundles over M is algebraically realized if and only if the set of all G vector bundles over M^τ is algebraically realized.*

As one might expect, one can utilize a result of Segal [Sg, Proposition 2.2] (use also the remark on the real case [Sg, p. 133–134]) and show fairly easily (compare [DM1, Proposition 7.2])

Proposition 5.5. *Let G be a compact Lie group. If M is a closed smooth manifold (with trivial action of G) then the set of all G vector bundles over M is algebraically realized.*

Example. Suppose G acts smoothly and semifreely, but not freely on a closed smooth manifold M. Then G acts freely on the sphere in a normal slice of a fixed point, and it follows that every element of order 2 in G is central [Mi3]. Assuming G is not of odd order we deduce Theorem B under the assumptions of the discussion from the previous two proposition.

6. Suggestions for Future Work

Apparently, there is still a long way to go from the results we have obtained so far to the conjectures we proposed. The most desirable way to close this gap could be a proof of Tognoli's theorem which is independent of the bordism calculation, i.e., a "natural proof" as Palais would call it. Such proofs have been attempted, but so far they have been unsuccessful, and there is little hope to confirm the algebraic realization conjecture this way.

On the other hand, the calculation of the equivariant bordism ring is available only in special cases. But, we have developed sufficiently many tools in real algebraic transformation groups to turn almost every available explicit calculation into a positive answer to the algebraic realization conjecture. Some progress still appears possible based on the computational approach to equivariant bordism, particularly for the bundle realization problem. One could use the methods in [Ko] and apply them in the calculation of the equivariant bordism of the Grassmannians.

More tools are needed to fully exploit the results on the equivariant bordism ring which are more structural [St1]. E.g., in this reference Stong proves a splitting theorem which reduces the calculation to relative bordism groups of families of 2-groups [St1, p. 28]. He also tries to reduce the problem of calculating the bordism to the case where the group is a 2 group, which does not work completely. Never-the-less, he states "*In particular, any property of the bordism for* 2 *groups which is inherited by direct summands and which holds for direct sums in which each summand has the property will hold for all equivariant bordism groups.*", see [St1, p. 36]. With this information on equivariant bordism, it seems interesting to develop sufficient machinary to show

Suggestion 6.1. *Let G be a finite group. Every closed smooth G manifold is algebraically realized if every closed smooth G_2 manifold is algebraically realized (here G_2 denotes the Sylow* 2 *subgroup of G).*

Typically, equivariant bordism calculations have the goal to break up the calculations into pieces, each of which is, preferably, a non-equivariant bordism problem. But, the subproblems are bordism problems of certain classifying spaces, even in case one starts out with the equivariant bordism problem of a point. Thus one is faced with task of showing that these classifying spaces have totally algebraic bordism, or homology, in the sense discussed in the previous section. Unfortunately, all too few spaces are known to have this property, the typical effort in the literature is to show that spaces don't have totally algebraic homology. Using homogeneous spaces as finite approximations for these classifying spaces one gets at least a candidate for the algebraic structure to work with (use Theorem D). Currently, there are still few tools around to show that these classifying spaces have totally algebraic homology.

Let us view the problem from a different perspective. Let G be a compact Lie group. Impose a canonical algebraic structure on the finite approximation of BG, the classifying space of G. Here canonical means that the universal principal bundle should be algebraic, in an appropriate sense. Then we can study strongly algebraic principal G fibrations, i.e., fibrations with entire rational classifying map. If finite approximations of BG have totally algebraic homology, then Theorem 4.1 implies that a principal G fibration over a closed smooth manifold is algebraically realized. (It would be desirable to even show that the set of all principal G fibrations over the manifold is algebraically realized, the definitions should be analogous to the ones for vector bundles, see Definition 2.2. Many techniques for such a result are not in place.) Using methods as in the proof of Proposition 4.3, one is then able to conclude that free actions of G on closed smooth G manifolds are algebraically realized. True, we know that this is the case. Still, in an attempt to rely less on equivariant bordism results, it is desirable to have a approach which provides an alternative argument. Even more important is the question whether the argument can be reversed to use Theorem A (3) to show that finite approximations of BG have totally algebraic homology.

Many constructions for smooth manifolds, such as the cartesian product and disjoint union, can also be applied in the real algebraic category, but the orbit space construction, which is often used in the description of bordism results, is an exception. The quotient of a free action on a variety is a semi-algebraic set. This problem has been studied quite carefully in [Sw3]. Commonly, these orbit spaces are balanced products, and topologically one has a fibration $H \to G \times X \xrightarrow{\pi} G \times_H X$. This relates our problem to the ideas expressed in the previous paragraph, though they need to be explored G equivariantly. If H has finite index in G, then the problem has been addressed in the induction result, stated as Proposition 4.5. If X is a point, then it is completely understood, see Theorem D. So, there is hope.

Let us pass from the existence question of an algebraic structure on $G \times_H X$ to the uniqueness question. As we have seen in Theorem D, the assumption that the orbit map π is algebraic (or entire rational) relates the algebraic structures of the total and base space of the fibration π. Which assumptions assure that they determine each other uniquely?[5] This problem is open also if H has finite index in G. A good understanding may give us valuable insight.

In his thesis, Ozan considers the case where the total space and the base space of an orbit map have incompatible real algebraic structures [Oz]. Let G be a finite cyclic group. Interpreted in our context, Ozan shows that one may impose real algebraic structures X on closed smooth manifolds such that not every homotopy class of maps from X to BG has an entire rational representative.

Previously, we considered a different approach to the fixed point conjecture which would avoid the use of closed manifolds [DS2]. The following suggested theorem has the fixed point conjecture as an immediate consequence. It also includes the algebraic realization conjecture and generalizes the non-equivariant result of Akbulut and King quoted as Theorem 1.3, and is particularly appealing as it tells us exactly what real algebraic G varieties are.

Suggestion 6.2. *Up the diffeomorphism, the set of non-singular real algebraic G varieties is exactly the set of interiors of compact smooth G manifolds with (possibly empty) boundaries.*

---

[5]Ed Bierstone raised this question after the author's lecture during the real algebraic geometry at Michigan State University in December of 1993

At the present time, a proof, though quite technical, seems within reach only for groups of odd order. The strategy would be once again to follow the non-equivariant approach of Akbulut and King in [AK1]. First of all, one needs a relative version of Theorem 4.1, and as King remarked, this is typically true if one has the definitions right (see the general principle in Section 1). Combining this with a relative equivariant bordism calculation one expects a relative version of Tognoli's theorem. Then one uses again the ideas of Akbulut and King to reduce the problem to one for closed manifolds. In the reduction, one needs to get rid of fixed point components which do not extend to the boundary. This is possible for odd order groups through a process which one may call topological blow up. Through a tightly controlled geometric process or a calculation, one may have to deal with equivariant Whitehead torsion obstruction as one needs to recognize certain complementary disks as disks in linear representations. Anyway, this project is quite involved and technical, and the application it was designed to prove has been solved completed differently.

REFERENCES

[AK1] S. Akbulut and H. King, *The Topology of Real Algebraic Sets with Isolated Singularities*, Annals of Math. **113** (1981), 425–446.

[AK2] S. Akbulut and H. King, *A Relative Nash Theorem*, Trans. of the Amer. Math. Soc. **267** (1981), 465–481.

[AK3] S. Akbulut and H. King, *Some New Results on the Topology of Nonsingular Real Algebraic Sets*, Bull. Amer. Math. Soc. **23** (1990), 441–446.

[AK4] S. Akbulut and H. King, *On Approximating Submanifolds by Algebraic Sets and a Solution to the Nash Conjecture*, Invent. Math. **107** (1992), 87–89.

[AK5] S. Akbulut and H. King, *The Topology of Real Algebraic Sets*, M.S.R.I. Publications no. 25, Springer Verlag, New York, Berlin, Heidelberg, London, Paris, Tokyo, Hong Kong, Barcelona, Budapest, 1992.

[AK6] S. Akbulut and H. King, *All Compact Manifolds are Homeomorphic to Totally Algebraic Real Algebraic Sets*, Comment. Math. Helv. **66** (1991), 139–149.

[BM] A. Bak and M. Morimoto, *Equivariant Surgery on Compact Manifolds with Half Dimensional Singular Sets*, preprint 1991.

[BD] R. Benedetti and M. Dedò, *Counterexamples to Representing Homology Classes by Real Algebraic Subvarieties up to Homeomorphism*, Compositio Math. **53** (1984), 143–151.

[BT] R. Benedetti and A. Tognoli, *On Real Algebraic Vector Bundles*, Bull. Sci. Math. **104** (1980), 89–112.

[Bi] D. Birkes, *Orbits on Linear Algebraic Groups*, Ann. of Math. **93** (1971), 459–475.

[BCR] J. Bochnak, M. Coste, and M.-F. Roy, *Géometrie Algébraique Réele*, Ergeb. der Math., Springer Verlag, Berlin, Heidelberg, New York, 1987.

[BK] J. Bochnak and W. Kucharz, *Nonisomorphic Algebraic Models of a Smooth Manifold*, Math. Ann. **290** (1991), 1–2.

[Bo] A. Borel, *Linear Algebraic Groups (Second Edition)*, Graduate Text in Mathematics, Vol. 126, Springer Verlag, New York, Berlin, Heidelberg, London, Paris, Tokyo, Hong Kong, Barcelona, 1991.

[BH] A. Borel and Harish-Chandra, *Arithmetic Subgroups of Algebraic Groups*, Ann. of Math. **75** (1962), 485–535.

[Br] G. Bredon, *Introduction to Compact Transformation Groups*, Academic Press, New York, 1972.

[CF] P. E. Conner and E. E. Floyd, *On the Construction of Periodic Maps without Fixed Points*, Proc. Amer. Math. Soc. **10** (1959), 354–360.

[CM] P. E. Conner and D. Montgomery, *An Example for $SO(3)$*, Proc. Nat. Acad. Sci. U. S. A. **48** (1962), 1918–1922.

[Co] S. R. Costenoble, *Unoriented Bordism for Odd-Order Groups*, Topology and its Applications **28** (1988), 277–287.

[DKS] K. H. Dovermann, F. Knop, and D. Y. Suh, *Topological Invariants of Real Algebraic Actions*, Topology and its Applications **40** (1991), 171–188.

[DM1] K. H. Dovermann and M. Masuda, *Algebraic Realization of Manifolds with Group Actions*, (to appear in Advances in Mathematics).

[DM2] K. H. Dovermann and M. Masuda, *Algebraic Realization of Closed Smooth Manifolds with Cyclic Actions*, (preprint) (1993).

[DM3] K. H. Dovermann and M. Masuda, *Fixed Point Free Low Dimensional Real Algebraic Actions of A_5 on Contractible Varieties*, Comment. Math. Helvetici **65** (1990), 534–546.

[DMP] K. H. Dovermann, M. Masuda, and T. Petrie, *Fixed Point Free Algebraic Actions on Varieties Diffeomorphic to $\mathbb{R}^n$*, Topological Methods in Algebraic Transformation Groups (H. Kraft, T. Petrie, and G. Schwarz, eds.), Progress in Mathematics, Vol. 80, Birkhäuser, Boston, Basel, Berlin, 1989, pp. 49–80.

[DMS1] K. H. Dovermann, M. Masuda and D. Y. Suh, *Algebraic Realization of Equivariant Vector Bundles*, J. für reine und angewandte Mathematik, Crelle **448** (1994), 31–64.

[DMS2] K. H. Dovermann, M. Masuda and D. Y. Suh, *Nonisomorphic Algebraic Models of a Smooth Manifold with Group Action*, (to appear in Proc. of the A. M. S.).

[DS1] K. H. Dovermann and D. Y. Suh, *Smith Equivalence for Finite Abelian Groups*, Pacific J. Math. **152** (1992), 41–78.

[DS2] K. H. Dovermann and D. Y. Suh, *Real Algebraic Transformation Groups*, Topological and Geometric Structures of Manifolds (Yong-Seung Cho, ed.), Proceedings of Workshops in Pure Mathematics, Mokpo National University (1991) Vol. 11, Part III, Pure Mathematics Research Association, The Korean Academic Council, pp. 60–105.

[EL] A. L. Edmonds and R. Lee, *Compact Lie Group which Act on Euclidean Space without Fixed Points*, Proc. Amer. Math. Soc. **55** (1976), 416–418.

[FR] E. E. Floyd and R. W. Richardson, *An Action of a Finite Group on an n-Cell without Stationary Points*, Bull. Amer. Math. Soc. **65** (1959), 73–76.

[Ho] G. Hochschild, *The Structure of Lie Groups*, Holden-Day, San Francisco, London, Amsterdam, 1965.

[Hi] M. W. Hirsch, *Differential Topology*, Graduate Texts in Mathematics Vol. 33, Springer Verlag, New York, Heidelberg, Berlin, 1976.

[Hu] J. E. Humphreys, *Linear Algebraic Groups*, GTM Vol. 21, Springer Verlag, New York, Heidelberg, Berlin, 1975.

[HH] W. C. Hsiang and W. Y. Hsiang, *Differentiable Actions of Connected Classical Groups, I*, Amer. J. of Math. **89** (1967), 705–786.

[Iv] N. V. Ivanov, *Approximation of Smooth Manifolds by Real Algebraic Sets*, Russian Math. Surveys **37:1** (1982), 1–59.

[Ka] T. Kambayashi, *Automorphism Group of a Polynomial Ring and Algebraic Group Actions on an Affine Space*, Jour. of Algebra **60** (1979), 439–451.

[KM] S. S. Kim and M. Masuda, *Topological Characterizations of Non-singular Real Algebraic G-Surfaces*, Topology and its Applications **57** (1994), 31–39.

[Kg] H. King, *Topological Problems Arising from Real Algebraic Geometry*, Topology Hawaii (K. H. Dovermann, ed.), World Scientific Publishing, Singapore, 1991, pp. 149–156.

[Kn] F Knop, *Nichtlinearisierbare Operationen halbeinfacher Gruppen aud affinen Räumen*, Invent. Math. **105** (1991), 217–220.

[Ko] C. Kosniowski, *Actions of Finite Abelian Groups*, Pitman, London, San Francisco, Melbourne, 1978.

[Kr1] H. Kraft, *Geometrische Methoden in der Invariantentheorie*, Friedrich Vieweg & Sohn, 1985.

[Kr2] H. Kraft, *G-vector bundles and the linearization problem*, Group actions and invariant theory, CMS Conference Proceedings vol. 10, 1989, pp. 111–123.

[KS] H. Kraft and G. Schwarz, *Reductive Group Actions with One-dimensional Quotient*, Inst. Hautes Études Sci. Publ. Math. **76** (1992), 1–97.

[LM] E. Laitinen and M. Morimoto, *Finite Groups with Smooth One Fixed Point Actions on Spheres*, preprint 1993.

[Ma] M. Masuda, *Algebraic Transformation Groups from a Topological Point of View*, Sugaku Expositions **6** (1993), 125–146.

[MP] M. Masuda and T. Petrie, *Equivariant Algebraic Vector Bundles over Representations of Reductive Groups: I*, Proc. Natl. Acad. Sci. USA **88** (1991), 9061–9064.

[MMP] M. Masuda, L. Moser-Jauslin, and T. Petrie, *Equivariant Algebraic Vector Bundles over Representations of Reductive Groups: II*, Proc. Natl. Acad. Sci. USA **88** (1991), 9065–9066.

[Mk] H. Miki, *Non-linearizable Compact Real Algebraic Group Actions on* $\mathbb{R}^n$, Master's Thesis, Osaka City Univ. (in Japanese) (1991).

[Mi1] J. Milnor, *On the Stiefel-Whitney Numbers of Complex Manifolds and of spin-Manifolds*, Topology **3** (1965), 223–230.

[Mi2] J. Milnor, *Singular Points of Complex Hypersurfaces*, Ann. of Mathematics Studies, Vol. 61, Princeton University Press, Princeton, N. J., 1968.

[Mi3] J. Milnor, *Groups which Act on* S^n *without Fixed Points*, Amer. J. Math. **79** (1957), 623–630.

[Na] J. Nash, *Real Algebraic Manifolds*, Annals of Math. **56** (1952), 405–421.

[O] R. Oliver, *Fixed-Point Sets of Group Actions on Finite Acyclic Complexes*, Comment. Math. Helvetici **50** (1975), 155–177.

[Oz] Y. Ozan, *Ph.D. Thesis, Michigan State University* (1994).

[Pa] R. Palais, *Equivariant, Real Algebraic Differential Topology*, preprint (1972).

[P1] T. Petrie, *Smooth Conjugacy of Algebraic Actions on Affine Varieties*, Topology **27** (1988), 473–477.

[PR1] T. Petrie and J. Randall, *Finite-order Algebraic Automorphisms of Affine Varieties*, Comment. Math. Helvetici **61** (1986), 203–221.

[PR2] T. Petrie and J. Randall, *A Topological View of Algebraic Actions on Complex Affine Space*, Proc. of KIT Math. Workshop, Algebra and Topology 1986 (J. K. Koo and D. Y. Suh, eds.), Korea Institute of Technology, Mathematics Research Center, 1987, pp. 272–292.

[Sg] G. Segal, *Equivariant K-Theory*, Inst. Hautes Études Sci. Publ. Math. **34** (1968), 129–151.

[Sf] H. Seifert, *Algebraische Approximation von Mannigfaltigkeiten*, Math. Zeitschrift **41** (1936), 1–17.

[Sw1] G. Schwarz, *Smooth Functions Invariant under the Action of a Compact Lie Group*, Topology **14** (1975), 63–68.

[Sw2] G. Schwarz, *Exotic Algebraic Group Actions*, C. R. Acad. Sci. Paris **309** (1989), 89–94.

[Sw3] G. Schwarz, *The Topology of Real Algebraic Quotients*, Topological Methods in Algebraic Transformation Groups (H. Kraft, T. Petrie, and G. Schwarz, eds.), Progress in Mathematics, Vol. 80, Birkhäuser, Boston, Basel, Berlin, 1989, pp. 135–152.

[St1] R. E. Stong, *Unoriented Bordism and Actions of Finite Groups*, Memoirs of the Amer. Math. Soc. **103** (1970).

[St2] R. E. Stong, *Equivariant Bordism and* $(\mathbb{Z}_2)^k$ *Actions*, Duke J. of Math. **37** (1970), 779–785.

[To] A. Tognoli, *Su una Congettura di Nash*, Annali della Scuola Normale Superiore di Pisa **27** (1973), 167–185.

[W] H. Whitney, *Elementary Algebraic Structures of Real Algebraic Varieties*, Ann. of Math. **66** (1957), 545–556.

Department of Mathematics, University of Hawaii, Honolulu, HI 96822, USA
E-mail address: dover@ uhunix.bitnet

Contemporary Mathematics
Volume **182**, 1995

ONE FIXED POINT ACTIONS ON SPHERES (APPENDIX)

KARL HEINZ DOVERMANN

ABSTRACT. We show that groups of odd order, simple groups, and nilpotent groups act on a disk without a fixed point if and only if they act smoothly on a sphere with exactly one fixed point. More generally, this statement has been shown for an arbitrary finite group by Laitinen and Morimoto, but important, rather delicate parts of the argument due to Bak and Morimoto are still unavailable. As we are using this result in connection with the fixed point conjecture, we include the argument.

A.0 Introduction

A finite group G is called an *Oliver group* (this expression was coined by Laitinen and Morimoto) if there does not exist a normal series $P \lhd H \lhd G$ such that P and G/H are of prime power order, and H/P is cyclic. (The trivial group is considered to be of prime power order.) Oliver proved that a finite group G acts smoothly (or simplicially) on a disk without fixed point if and only if G is an Oliver group [O1]. If G acts smoothly on a sphere with exactly one fixed point, then it also has an action on a disk without fixed point, and it follows that G is an Oliver group. The converse is a long standing problem in transformation groups. In 1946 Montgomery and Samelson asked whether there are actions on spheres with exactly one fixed point [MS]. The first actions of this kind were constructed by Stein for $G = \mathrm{SL}(2, \mathbb{Z}_5)$ and $G = \mathrm{SL}(2, \mathbb{Z}_5) \times \mathbb{Z}_r$ where $(r, 120) = 1$. Soon after this, Petrie showed that abelian Oliver groups of odd order (i.e., groups with at least 3 non-cyclic Sylow subgroups) and some other finite and infinite groups, among them A_5, can act smoothly on a (homotopy[1]) sphere with exactly one fixed point [P1], [P2]. He proposed to determine all groups which can act on a homotopy sphere with exactly one fixed point [P3, p. 183]. Among experts, the suggested solution is that these groups are exactly the ones which can act on a disk without fixed point, i.e., the Oliver groups. Pawalowski stated this as a conjecture in [Pa, p. 173]. A confirmation of this conjecture for non-solvable groups was provided by Laitinen, Morimoto, and Pawalowski [LMP], and this work was extended by Laitinen and Morimoto to confirm the conjecture for any finite group [LM].

Theorem. *A finite group has a smooth action with exactly one fixed point on a standard sphere if and only if it is an Oliver group.*

1991 *Mathematics Subject Classification.* 57S17, 57S25.

[1]Some of the literature on this subject emphasizes actions on standard spheres, some contents itself with actions on homotopy spheres. As it turns out, due to a fairly simple construction, these two problems are essentially equivalent (see Lemma A.12.2).

Both papers, [LMP] and [LM], depend on delicate surgery theoretic work of Bak and Morimoto [BM], and so far the latter one has not been available. On the other hand, we have shown the algebraic realization theorem only for certain classes of groups, so that we could make use of the theorem of Laitinen and Morimoto only for these groups. In many of these cases we do not need the results of Bak and Morimoto, but instead we can combine results from [LM] and known methods in equivariant surgery theory to prove the existence of one fixed point actions on spheres. Specifically, we show the theorem for odd order groups, simple groups, and nilpotent groups, or more generally for groups which have an admissable representation which satisfies the strong gap condition (see A.2.5).

In the same context, one may try to find the lowest dimension of a sphere on which there is an action with exactly one fixed point. Morimoto showed that A_5 acts on the standard sphere in dimension 6 with exactly one fixed point [Mo1], and this is the lowest dimension in which any group has a one fixed point action on a sphere. (See [Fu], [Mo2], [DeM] and [BKS], and the general discussion in [BKS].)

Many ideas from the papers [LMP] and [LM] enter into our version. On the other hand, we make an effort to use only published results as references, and if they are not available, include details. There is one exception, in Sections 1 and 2 we quote results from Laitinen and Morimoto, but these are proved in detail in [LM]. In the latter part of this appendix the reader will find few references to [LM]. The reason for this is that their and our discussion have followed different paths. Still, we learned several key ideas from their paper, and they entered in our proof. In broad terms, one may say that the proof in [LM] specializes to ours as the surgery obstruction groups of [BM] specialize to the Wall groups we are using.

A.1 Facts about the Burnside Ring

Let G be a finite group and p a prime, then G^p denotes the smallest normal subgroup of G such that G/G^p is a group of p power order. It is well defined. Denote by $\mathcal{S}(G)$ the set of all subgroups of G. It is a G set with G acting by conjugation. Set

$$\mathcal{L}(G) = \{H \in \mathcal{S}(G) \mid H \supseteq G^p \text{ for some prime } p\} \text{ and } \mathcal{M}(G) = \mathcal{S}(G) \setminus \mathcal{L}(G).$$

Both sets $\mathcal{L}(G)$ and $\mathcal{M}(G)$, are G invariant, and $\mathcal{L}(G)$ is closed in $\mathcal{S}(G)$ in the sense that whenever $H \in \mathcal{L}(G)$, $K \in \mathcal{S}(G)$, and $H \subset K$, then $K \in \mathcal{L}(G)$.

Let $\Omega(G)$ denote the Burnside ring of G [Dr1], [tD1]. For every subgroup H of G we have a ring homomorphism

$$\chi_H : \Omega(G) \to \mathbb{Z} \text{ defined by } \chi_H([X]) = \chi(X^H).$$

Here X is a finite G CW complex, $[X]$ its class in $\Omega(G)$, X^H its H fixed point set, and χ the Euler characteristic. Let $\mathrm{Res}^G_H : \Omega(G) \to \Omega(H)$ denote the homomorphism obtained by restricting G actions to H actions. A reformulation of the results of Section 1 of [LM] is as follows.

Theorem A.1.1. *Let G be a finite group. There exists an element $\beta \in \Omega(G)$ such that $\chi_G(\beta) = 1$ and $\mathrm{Res}^G_H(\beta) = 0$ for all $H \in \mathcal{M}(G)$.*

More precisely, Laitinen and Morimoto show that $\mathcal{M}(G)$ is the (unique, open) maximal subset of $\mathcal{S}(G)$ for which there exists an element $\alpha \in \Omega(G)$ with the properties stated in the theorem.

There is a variation of this theorem which will be useful later. Let $\mathcal{G}^1(G)$ denote the set of subgroups H of G for which there exists a short exact sequence $1 \to P \to H \to C \to 1$ where C is cyclic and P is of prime power order. Similarly, for two primes p and q we denote by $\mathcal{G}_p^q(G)$ the set of subgroups K of G for which there exists a normal series $P \lhd H \lhd K$ such that P is of p power order, H/P is cyclic, and K/H is of q power order.

Proposition A.1.2. *Let G be an Oliver group. There exists an element δ in $\Omega(G)$ such that $\operatorname{Res}_K^G \delta = -1$ for all $K \in \mathcal{G}^1(G)$, and if δ is expressed as a $\mathbb{Z}$ linear combination of homogeneous spaces, $\delta = \sum \delta_H[G/H]$, then $\chi_G(\delta) = 0$ and $\delta_H > 0$ if $H \in \mathcal{S}(G) \setminus \mathcal{G}^1(G)$ and $H \neq G$.*

Proof. There is a finite contractible G CW complex D without fixed point [O1]. We like to set $\delta = -[D]$. Then $\chi_K(D) = 1$, hence $\operatorname{Res}_K^G(D) = 1$ and $\operatorname{Res}_K^G(\delta) = -1$, for all $K \in \mathcal{G}^1(G)$. Furthermore, $\chi_G(\delta) = \chi_G[D] = 0$. For such an action $\chi_H(D)$ is arbitrary if $H \notin \mathcal{G}_q^p(G)$ for any primes p and q, and determined up to a congruence if $H \in \mathcal{G}_q^p(G)$ for some primes p and q but $H \notin \mathcal{G}^1(G)$, see [O1]. Hence we may prescribe $\chi_H(D)$ for $H \neq G$ and $H \notin \mathcal{G}^1(G)$ such that $\delta_H > 0$ for all H as in our claim. □

A.2 A Representation of G

Let $\mathbb{R}[G]$ denote the regular representation of G and $\mathbb{R}$ the trivial representation of G. If H is a subgroup of G, then $\mathbb{R}[G]^H$ may be identified with the permutation representation $\mathbb{R}[H\backslash G]$ on the set of right cosets $H\backslash G$. With this and the fact that right and left cosets are in equivariant 1-1 correspondence if H is a normal subgroup of G, Laitinen and Morimoto define

$$W := (\mathbb{R}[G] - \mathbb{R}) - \bigoplus_p (\mathbb{R}[G/G^p] - \mathbb{R}) \tag{A.2.1}$$

where p ranges over all primes and the minus sign is understood representation theoretically. It takes some checking to see that the direct sum is a summand of $\mathbb{R}[G] - \mathbb{R}$. If H and K are subgroups of G, then we denote by $\langle H, K\rangle$ the subgroup of G generated by H and K. If either H or K is a normal subgroup of G, then $\langle H, K\rangle = HK$. For any subgroup H of G (count orbits)

$$\dim W^H = ([G:H] - 1) - \sum_p ([G : HG^p] - 1). \tag{A.2.2}$$

Let $\mathcal{P}(G)$ denote the set of subgroups of G of prime power order. As a reformulation of Theorems 2.3 and 3.2 of [LM] we state

Theorem A.2.3. *Let G be a finite group and W as in A.2.1. Then*

(a) *$W^H \neq 0$ if and only if $H \in \mathcal{M}(G)$.*
(b) *$\dim W^H \geq [K:H] \dim W^K$ if $H \subseteq K \in \mathcal{S}(G)$.*
(c) *Suppose $H \in \mathcal{M}(G)$ and $H \subseteq K \in \mathcal{S}(G)$. Then $\dim W^H = 2 \dim W^K$ if and only if $[K:H] = 2$, $[KG^2 : HG^2] = 2$, and $HG^p = G$ for all odd primes p.*
(d) *If $\dim W^P = 2 \dim W^{H_i}$ for $H_i \in \mathcal{S}(G)$, $i = 1, 2$, and $P \in \mathcal{P}(G)$ where $P \subset H_i$, then $\langle H_1, H_2\rangle \in \mathcal{M}(G)$.*

Let U be a representation of G with G invariant inner product and $H \in \mathcal{S}(G)$. Then U decomposes as a direct sum $U^H \oplus U_H$, where U_H denotes the orthogonal complement of U^H in U. A collection of orientations of positive dimensional U^H and U_H for all $H \in \mathcal{S}(G)$ is called a *G-orientation* of U. In particular, realifications of complex representations of G have naturally defined G-orientations.

Definition A.2.4. *A real representation V of G with G invariant inner product is called admissable if it is the realification of a complex representation of G (and as such we consider it as a G-oriented representation) such that*

(a) $\dim V^H \geq 6$ *for all* $H \in \mathcal{G}^1(G)$.
(b) *If* $V^H \neq 0$ *then* $\dim V^H \geq 4$ *and* $H \in \mathcal{M}(G)$.
(c) $\dim V^H \geq [K : H] \dim V^K$ *if* $H \subseteq K \in \mathcal{S}(G)$.
(d) *If* $\dim V^P = 2 \dim V^{H_i}$ *for* $H_i \in \mathcal{S}(G)$, $i = 1, 2$, *and* $P \in \mathcal{P}(G)$ *with* $P \subset H_i$, *then* $[H_i : P] = 2$ *and* $\langle H_1, H_2 \rangle \in \mathcal{M}(G)$.

For an Oliver group one observes that $\mathcal{G}^1(G) \subseteq \mathcal{M}(G)$ [LM, Proposition 1.2]. By definition we also have that $\mathcal{P}(G) \subset \mathcal{G}^1(G)$. Let us say furthermore that G is 2-perfect if $G^2 = G$.

Theorem A.2.5. *Every Oliver group has admissable representations. More specifically, if G is an Oliver group, l an even integer ≥ 6, and W as in A.2.1, then $V = lW$ is an admissable representations, and its dimension is*

$$n = l\{(|G| - 1) - \sum_p ([G : G^p] - 1)\}.$$

The sum ranges over all primes which divide the order of G. Furthermore, if G is 2-perfect (e.g., if G is of odd order or simple) or nilpotent (e.g., if G is abelian), then V satisfies the strong gap condition

(SGC) $\dim V^H > 2(\dim V^K + 1)$ *for* $H \in \mathcal{M}(G)$, $H \subseteq K \in \mathcal{S}(G)$, *with* $H \neq K$.

Proof. The first claim is a consequence of A.2.3 and has been verified by Laitinen and Morimoto, see [LM, Proposition 3.4]. We verify the second claim.

Appealing to Theorem A.2.3 (c), we need to check that, under the given assumptions on G, there are no subgroups H, $K \in \mathcal{S}(G)$ such that $H \subset K$, $H \in \mathcal{M}(G)$, $[K : H] = 2$, $[KG^2 : HG^2] = 2$, and $HG^p = G$ for all odd primes p. If G is 2-perfect, then we always have that $[KG^2 : HG^2] = 1$. If G is nilpotent and $HG^p = G$ for all odd primes p, then $H \supset G^2$ and $H \notin \mathcal{M}(G)$. Hence, under either assumption on G, it does not happen that $\dim V^H = 2 \dim V^K$, and the strong gap condition will be satisfied by V. □

A.3 G Manifolds with V Structure

We discuss the bundle date, called V structure as in [P1], for the manifolds to which we are going apply equivariant surgery.

Let K be a finite group, U and V K-oriented representations of K, and $L : U \to V$ an equivariant isomorphism. Then L induces isomorphisms $L^H : U^H \to V^H$ and

$L_H : U_H \to V_H$. We say that L is K-orientation preserving (reversing) if, for all subgroups H of K, L^H is orientation preserving (reversing) and L_H is orientation preserving whenever U^H and U_H are positive dimensional. In case U_H is zero dimensional, we only require that L^H is orientation preserving (reversing). In case U^H is zero dimensional, we require that L_H is orientation preserving (reversing). A smooth G manifold X is said to be *G-oriented*[2] if each positive dimensional component of X^H is oriented for each subgroup H of G. If X is a G oriented manifold, and $x \in X$ is a point with isotropy group H, then the tangent space of X at x, T_xX, has a naturally defined H-orientation. One just needs to observe that for every subgroup K of H we have a direct sum decomposition $T_xX = (T_xX)^K \oplus (T_xX)_K$, that, by assumption, T_xX and $(T_xX)^K = T_xX^K$ are oriented, and that there is a unique orientation of $(T_xX)_K$ so that the direct sum decomposition preserves orientations. If V is a representation of G, then $\underline{V}$ denotes a product bundle with fibre V whose base space is understood from context.

Next we recall the definition of a V structure. In comparison to [P1], where this concept was introduced, we will also be specific about G orientations to the extent that they are needed in Lemma A.3.2. Given a G-vector bundle ξ over X, we obtain a collection of N_GH-vector bundles $\{(\xi_{|X^H})_H \mid H \in \mathcal{S}(G)\}$, we restrict ξ over X^H and take the orthogonal complement of the H-fixed point set. This collection is denoted by $\pi(\xi)$, and as the indexing set is $\mathcal{S}(G)$, we call it an $\mathcal{S}(G)$ bundle system. If X is a smooth G manifold and $\xi = TX$, then $\pi(\xi) = \{\nu(X^H, X)\}$ is the collection of normal bundles of X^H in X, also denoted by $\nu(X)$, and if $\xi = \underline{V}$, then $\pi(\xi) = \{\underline{V_H} \mid H \in \mathcal{S}(G)\}$. There is a general notion of π bundles [DP1], but we only need the two examples given above.

Let ξ and η be G vector bundles over X and $\varphi : \xi \to \eta$ a G vector bundle isomorphism. The collection $\pi(\varphi) = \{\varphi_H : (\xi_{|X^H})_H \to (\eta_{|X^H})_H \mid H \in \mathcal{S}(G)\}$ provides an isomorphism of the bundle systems $\pi(\xi)$ and $\pi(\eta)$. Each φ_H is an N_GH vector bundle isomorphism, and there are certain natural compatibilities. One takes these properties of the system $\{\varphi_H \mid H \in \mathcal{S}(G)\}$ as the guiding principle, without requiring that the individual φ_H are induced from a bundle isomorphism φ, to give a general definition of an isomorphism of bundle systems [DP1].

Next we incorporate the idea of orientations. Let X be a smooth G manifold and V a representation of G. Suppose both of them are G oriented. Then $\nu(X)$ and $\pi(\underline{V})$ are oriented bundle systems in the sense that each bundle in these collections of bundles is oriented.

We may stabilize an (oriented) G vector bundle isomorphism $\varphi : \xi \to \eta$ by adding the same (oriented) vector bundle μ to each of them and defining $\varphi \oplus \mathrm{Id} : \xi \oplus \mu \to \eta \oplus$

[2]It is sensible, though not crucial, to require some compatibilities for the orientations on X^H and X^K if $X^H = X^K$ or if H and K are conjugate to each other. If one likes to orient X^H for all $H \in \mathcal{S}(G)$, then one should proceed as suggested in [L, §2] or [tD2, p. 169]. In equivariant surgery theory, it is more convenient to only impose orientations on tangent spaces, and do so only on isotropy components [DP1]. Bundle data will take care of the normal directions, at least if one does surgery on positive dimensional fixed point components. For surgery on zero dimensional fixed point components we establish special conventions in A.3.5. We prefer to not have to address compatibilities of orientations explicitly, as only little is required in the special setting of our paper, and note that sufficient compatibilites for surgery on positive dimensional H fixed point components ($H \in \mathcal{S}(G)$) are imposed in A.3.1 (4). Using an admissable representation V as a model, see A.6.1, makes the choices of orientations not only compatible, but also natural.

μ as the stabilized (oriented) isomorphism. This generates an equivalence relation whose classes are called stable (oriented) bundle isomorphisms. An isomorphism $c = \{c_H \mid H \in \mathcal{S}(G)\}$ of (oriented) bundle systems is stabilized by stabilizing each c_H individually. The stable isomorphism of bundles (or bundle systems) is then denoted by $s(\cdot)$.

Definition A.3.1. *Let X be a smooth G manifold and V a representation of G. A V structure on X consists of a pair (b, c) where*

(1) *$b : TX \to \underline{V}$ is a stable G vector bundle isomorphism,*
(2) *$c : \nu(X) \to \pi(\underline{V})$ is an isomorphism of $\mathcal{S}(G)$ bundle systems, and*
(3) *$s(c) = \pi(b)$.*

Suppose that X and V are G oriented. Then we say that the V structure (b, c) preserves orientations in positive dimensions (popd) if for all $H \in \mathcal{S}(G)$

(4) *$b^H : TX^H \to \underline{V^H}$ is a stable oriented isomorphism whenever X^H is of positive dimension, and $c_H : \nu(X^H, X) \to \underline{V_H}$ is an oriented isomorphism.*

Caution. The assumption (popd) requires c_H to be orientation preserving even if X^H is of dimension zero.

The important observation is that a V structure provides adequate bundle data for equivariant surgery in positive dimensions as it is a special case of the bundle data considered in [DP1, §4]. There is an obvious notion of equivariant bordism for G manifolds with V structure (or G manifolds with V structure (popd)) where it is assumed that the V structures (including the orientations on fixed point components in positive dimensions) extend over the bordism.

It will be convenient to assume that H fixed point sets are connected if they are of positive dimension. This can be arranged with the help of zero dimensional surgery [DP1, Lemma 9.1] under the assumption that the V structure preserves orientations in positive dimensions. In the reference the lemma is formulated for a setting where as additional data there is a map to a given space. In our setting this target space can be taken as a point.

Lemma A.3.2. *Let X be a G manifold and V an admissable representation of G (see A.2.4). Suppose both are G oriented and (b, c) is a V structure (popd) on X. Then $(X, (b, c))$ is equivariantly cobordant, as a G manifold with V structure (popd), to a G manifold with V structure (popd), $(X', (b', c'))$, such that $(X')^H$ is connected for all H with $\dim V^H > 0$ and $(X')^H = X^H$ for all H with $\dim V^H = 0$.*

Less stringent assumptions on the dimensions of V^H, $H \in \mathcal{S}(G)$, and their gaps would have been sufficient, but we will not need the lemma in greater generality than stated.

Remark A.3.3. Suppose that X is a G manifold with V structure (b, c) and a distinguished fixed point x_0, and V a G oriented representation. If X is G orientable and X^H is connected whenever X^H is of positive dimension, $H \in \mathcal{S}(G)$, then the G orientation on X may be chosen such that (b, c) preserves orientations in positive dimensions. To arrange orientations like this one just identifies the tangent space of X at x_0, $T_{x_0}X$, with V as a G oriented representation, providing a local G orientation on X at x_0 (popd). As X^H is assumed to be connected whenever

$\dim X^H$ is positive, it follows that the orientations are preserved over the entire positive dimensional components. If $x \in X^H$ and X^H is of dimension zero, then the orientation on $\nu(x, X)$ is chosen such that c_H, restricted over x, preserves orientations.

Conversely, we can also use the local G orientation on a neighbourhood of x_0, and with it the one on $T_{x_0}X$, to determine a G orientation on V by chosing an identification of T_xX with V.

Remark A.3.4. Suppose X is a G manifold, V is a representation of G, and both are G oriented. Suppose (b, c) is a V structure on X which preserves orientations in positive dimensions. We may define $-X$. It is a G oriented manifold, the underlying G manifold is the same as X, but the orientation on $(-X)^H$ is taken as the reversed orientation of the one of X^H whenever X^H is of positive dimension. A V structure on $-X$ is defined by taking the same c as for X, but b is replaced by $(-b) = b \oplus (-\operatorname{Id}_{\mathbb{R}})$. So we added $\underline{\mathbb{R}}$ to the domain and range of b, and on this additional summand the map is multiplication with -1. It is trivial to observe that $(-b, c)$ defines a V structure on $-X$ which preserves orientations in positive dimensions.

Definition A.3.5. *Suppose X is a G manifold, V is a representation of G, and both are G oriented. Suppose (b, c) is a V structure on X which preserves orientations in positive dimensions. Let K be a subgroup of G such that $\dim V^K = 0$, and x_1, $x_2 \in X$ points with isotropy group K. Denote the restriction of the stable isomorphism b^K over x_i by b_i^K. We call (x_1, x_2) a cancelling pair if x_1 and x_2 lie in different G orbits, and one of the b_i^K is orientation preserving and the other one is orientation reversing.*

Example A.3.6. Let V be a G oriented representation of G such that $\dim V^G = 0$ and $Y = S(\mathbb{R} \oplus V)$. The natural isomorphism $\underline{\mathbb{R}} \oplus TS(\mathbb{R} \oplus V) \cong \underline{\mathbb{R}} \oplus \underline{V}$ provides a V structure on Y which preserves orientations in positive dimensions. The poles $(\pm 1, 0) \in Y$ form a cancelling pair.

Creating pairs of cancelling points. Let X be a G manifold with V structure (popd), and K a subgroup of G such that $\dim V^K = 0$. Then we can introduce a pair (x_1, x_2) of cancelling points with isotropy group K. One starts out with $X_1 = X \sqcup G \times_K S(\mathbb{R} \oplus V)$ and connects X_1^H whenever $\dim V^H$ is positive (Lemma A.3.2). The result is a G manifold with V structure (popd), and the poles in $S(\mathbb{R} \oplus V)$ form the newly created cancelling pair (see Example A.3.6).

Deleting pairs of cancelling points. Let X be a G manifold with V structure (popd), and (x_1, x_2) a pair of cancelling points with isotropy group $K \neq G$. By equivariant surgery of isotropy type K we can cancel this pair of points and obtain a G manifold X_1 with V structure (popd). Specifically, we attach $G \times_K (D^1 \times D(V))$ along $G \times_K (S^0 \times D(V))$ to $X \times I$ creating the trace W of the surgery. The core $K \times_K (D^1 \times \{0\})$ connects $x_1 \times \{1\}$ and $x_2 \times \{1\} \in X \times \{1\} \subset \partial(X \times I)$. The top end of W is X_1. The bundle data assumptions for a cancelling pair are made up to assure that W, and with it X_1, have a V structure (popd).

A.4 Remarks on Equivariant Surgery

Much of the literature on equivariant surgery is formulated for G normal maps

[DP1], [DR], [DS], [PR]. To simplify matters, we prefer to work with G manifolds X which have a V structure (b, c) in the sense of the previous section. This was also done in [P1]. Suppose X has a distinguished fixed point x_0. Set $Y = (T_{x_0}X)^+ = S(\mathbb{R} \oplus T_{x_0}X)$, where $(\cdot)^+$ denotes the one point compactification. We may identify a G invariant tubular disk neighbourhood D around x_0 with $T_{x_0}X$, and extend this identification to an equivariant map $f : X \to Y$ by mapping $X \setminus D$ to ∞ in Y. Under additional assumptions, which we specify later, $\mathcal{W} = (X, f, b, c)$ will be a G normal map, and when we talk about surgery on X, then this refers to surgery on $\mathcal{W}$. This is how our use of surgery is related to the one in the literature mentioned above.

A.5 One Fixed Point G Manifolds with V Structure

As a first approximation to the spheres with G action having exactly one fixed point we construct G manifolds with one fixed point. These will satisfy additional assumptions (e.g., they come equipped with a V structure), so that we my apply equivariant surgery to them. The ideas for the proof of our next theorem are taken from [P1].

Theorem A.5.1. *Suppose G is an Oliver group and V is an admissable representation of G. There exist closed smooth G manifold X with V structure and G orientation such that the action of G has exactly one fixed point and X^H is connected whenever it is of positive dimension.*

Proof. Let β be an element in the Burnside ring $\Omega(G)$, such that

$$\chi_G(\beta) = 0 \text{ and } \operatorname{Res}_H^G \beta = 1 \text{ in } \Omega(H) \text{ for all } H \in \mathcal{M}(G).$$

It exists by X.1.1. Set $\mu = 1 - \beta$. Then for every positive integer λ

$$\chi_G(1 - \mu^\lambda) = 0 \text{ and } \operatorname{Res}_H^G(1 - \mu^\lambda) = 1 \text{ in } \Omega(H) \text{ for all } H \in \mathcal{M}(G).$$

Let ω_G^0 denote the 0-th equivariant cohomotopy group, and let $\epsilon : \Omega(G) \to \omega_G^0(pt)$ be Segal's isomorphism [Se]. Specifically one has the following. Let U be a sufficiently large (e.g., suppose that U contains the regular representation of G), real, and G oriented representation of G. Given an element $\rho = \rho_+ - \rho_- \in \Omega(G)$, represented as a formal difference of finite G-sets ρ_+ and ρ_-, $\epsilon(\rho)$ may be taken as an equivariant proper selfmap of U which is transversal to $0 \in U$, such that $\epsilon(\rho)^{-1}(0) = \rho_+ \sqcup \rho_-$ and its derivative $d\epsilon(\rho)_{|x}$ is G orientation preserving if $x \in \rho_+$ and G orientation reversing if $x \in \rho_-$.

Now use that $\operatorname{Res}_H^G(1 - \mu^\lambda) = 1$ whenever $V^H \neq 0$. Set $Y := S(\mathbb{R} \oplus V)$ and let $y_\pm := (\pm 1, 0)$ be the north and south pole in Y. It follows from Petrie's localization result [P4, Lemma 4.16] that there exists an element $\omega \in \omega_G^0(Y)$ such that

$$\omega_- := (\omega_{|y_-}) = \epsilon(1) \text{ and } \omega_+ := (\omega_{|y_+}) = \epsilon(1 - \mu^\lambda)$$

for some positive integer λ. Specifically, we suppose that ω_- is the identity map, and $(\omega_+^{-1}(0))^G = \emptyset$.

By definition, ω may be represented as a quasi-equivalence $g : Y \times U \to Y \times U$ for sufficiently large representations U of G, i. e., g is proper, fibre preserving, and of degree 1 [P4, §7, p. 199 ff]. Modifying g by a G homotopy, we obtain an equivariant map $F : Y \times U \to Y \times U$ which is transverse to the zero section $Y \times \{0\}$ [P4, Corollary 5.4]. On $\{y_\pm\} \times U$ we may suppose that F equals g. Setting $X = F^{-1}(Y \times \{0\})$ we obtain a closed smooth G manifold, and the restriction of F to X defines an equivariant map $f : X \to Y$.

In Section 7 of [P4] Petrie shows how the stable tangent bundle data b is obtained from the transversality construction. The π-bundle isomorphism c with $\pi(b) = s(c)$ is constructed in [DP1, §8]. The orientability assumptions on X^H is discussed in [P4, §7]. More exactly, the transversality construction provides a stable isomorphism $\widetilde{b} : \underline{\mathbb{R}} \oplus TX \to f^*(\underline{\mathbb{R}} \oplus TY)$ and an isomorphism $\widetilde{c} : \nu(X) \to f^*(\nu(Y)) = f^*(\pi(\underline{\mathbb{R}} \oplus TY))$ of vector bundle systems such that each $\widetilde{b}^H$ and $\widetilde{c}_H$ is orientation preserving whenever X^H is of positive dimension. Compose $\widetilde{b}^H$ and $\widetilde{c}_H$ with the identifications $\underline{\mathbb{R}} \oplus TY \cong \underline{\mathbb{R}} \oplus \underline{V}$ and $\nu(Y) \cong \pi(\underline{V})$, which are popd, to obtain (b, c).

By construction, the action of G on X has exactly one fixed point, which lies in $F^{-1}(y_-, 0) = \omega_-^{-1}(0)$, and X is a G oriented manifold with V structure which preserves orientations in positive dimensions. We apply Lemma A.3.2 to assure that X^H is connected whenever it is of positive dimension. □

A.6 Geometric Modules over the Burnside Ring

The surgery obstruction groups are typically modules over the Burnside ring [Dr2]. The following geometric construction ($\ltimes$) is designed to have this multiplication as algebraic consequence. There is a substantial resemblence to the use of the additive properties of equivariant surgery obstructions [D1], [DP2], in particular the partial additions used in [D1]. Laitinen, Morimoto, and Pawalowski [LMP] carried this further by also using induction and restriction to incorporate the Burnside ring structure. In the context of this paper we apply the method to manifolds as we constructed them in Theorem A.5.1. We give them a name.

Definition A.6.1. *A closed smooth G manifold X with V structure (b, c) and distinguished base point is said to be special if*

(1) *V is admissable, X is G oriented, and (b, c) preserves orientations in positive dimensions,*
(2) *X^H is connected whenever it is of positive dimension, and*
(3) *X has a cancelling pair (x_1, x_2) with isotropy group $G_{x_i} = H$ for all $H \in \mathcal{S}(G)$, $H \neq G$, such that* $\dim V^H = 0$.

The Operations $\odot$ and $\ltimes$. Let X be a special G manifold with V structure (b, c), and $\alpha = \bigsqcup_H \alpha_H(G/H)$ a linear combination of homogeneous spaces where the α_H are integers and $\alpha_G = 0$. We construct a special G manifold with V structure which we denote by $(1+\alpha) \ltimes (X, (b, c))$, or $(1 + \alpha) \ltimes X$ for short if X is understood as a G manifold with V structure. First of all, set

$$(1 + \alpha) \odot X = X \sqcup \bigsqcup_H \alpha_H(G \times_H X).$$

In this formula, multiplication with α_H means that we take $|\alpha_H|$ copies of $G \times_H X$. We use the G orientation obtained from X if α_H is positive and its negative if α_H is negative. Denote the induced V structure on $G \times_H X$ by $G \times_H (b, c)$. Use the G orientation on X at x_0 (see Remark A.3.3) to define the G orientation on V. Then (b, c) and $G \times_H (b, c)$ preserve orientations in positive dimensions for X and $\bigsqcup \alpha_H(G \times_H X)$. This defines the V structure $(1+\alpha) \odot (b, c)$ on $(1 + \alpha) \odot X$. Obviously, $(1 + \alpha) \odot X$ has property (1) of a special G manifold.

We apply zero dimensional surgery to $(1 + \alpha) \odot (X, (b, c))$. Suppose for the moment that $\alpha = \pm(G/H)$. Observe that $(G \times_H X)^K$ consists of $|(G/H)^K|$ copies of X^K. Given $x_+ \in X^H$, the G orbit $[1, x_+] \in G \times_H X$ meets each copy of X^K exactly once. Pick a point x_- in X^H. If $\dim X^H = 0$, pick $([1, x_+], x_-)$ as a cancelling pair in $(1+\alpha) \odot X$. Imbed $G/H \times S^0$ equivariantly into $X \sqcup G \times_H X$ such that $\pm 1 \in S^0$ maps to $x_\pm$. Use this imbedding of $G/H \times S^0$ to perform equivariant surgery. If $\dim X^K$ is positive, then the surgery connects each component of $(G \times_H X)^K$ to X^K by exactly one (non-equivariant) zero dimensional surgery. So the K fixed point set of the resulting manifold is again connected. If the dimension of X^K is zero and $K \neq G$, then there will again be cancelling pairs of points of this isotropy type. If $K \neq H$, then the pair for K is unaffected by the surgery. If $H = K$, then we obtain the complementary points in the pairs of $[1, x_+]$ and x_- as new cancelling pair. Hence the resulting manifold again satisfies properties (2) and (3) of a special G manifold. Hence we obtain a special G manifold, which we denote by $(1 + \alpha) \ltimes X$, with an induced V structure, which we denote by $(1 + \alpha) \ltimes (b, c)$. Taken together we obtain $(1 + \alpha) \ltimes (X, (b, c))$.

For a general α we carry out the zero dimensional surgeries of the previous paragraph for each copy of $G \times_H X$ in $(1 + \alpha) \odot X$, one at a time.

Apparently the bordism class of $(1+\alpha) \ltimes (X, (b, c))$ depends only on the class of α in the Burnside ring and the bordism class of $(X, (b, c))$, though the diffeomorphism class depends on the specific choice of cancelling pairs along which we do surgery.

Suppose we embed $G \times_H S^k$ into a special G manifold X and use it to apply equivariant surgery. Denote the result by $\overline{X}$. Then (see [DP1, p. 51])

$$[\overline{X}] = [X] + (-1)^{k+1} 2[G/H] \quad \text{in} \quad \Omega(G). \tag{A.6.2}$$

Let $\alpha = \bigsqcup \alpha_H(G/H)$ be again an integral linear combination of homogeneous spaces such that $\alpha_G = 0$ and X a special G manifold. Set $|\alpha| = \bigsqcup |\alpha_H|(G/H)$. In $\Omega(G)$ we have

$$[(1 + \alpha) \ltimes X] = (1 + [|\alpha|])[X] - 2[|\alpha|]. \tag{A.6.3}$$

To see this equation consider the following identities.

$$[(1+\alpha) \ltimes X] = [(1+|\alpha|) \ltimes X] = [(1+|\alpha|) \odot X] - 2[|\alpha|] = (1+[|\alpha|])[X] - 2[|\alpha|]. \tag{A.6.4}$$

The first equality follows as the signs of the α_H determine only orientations and have no effect on Euler characteristics. The second equal sign follows from a count of the zero dimensional surgeries performed in the construction of $(1+|\alpha|) \ltimes X$ from $(1 + |\alpha|) \odot X$ together with A.6.2. The third equality follows from the observation that $1 + |\alpha|$ is a G set, and the fact that $G \times_H X$, with left G action, is equivalent as a G space to $(G/H) \times X$, with diagonal G action, see [Br, p. 113], noting that the latter operation defines the multiplication in $\Omega(G)$. $\square$

A.7 Inductive Assumptions

Let G be a finite group. A subset $\mathcal{K}$ of $\mathcal{S}(G)$ is said to be open in $\mathcal{S}(G)$ if it is the complement of a closed subset, compare subsection A.1. Recall the definition of $\mathcal{G}^1(G)$ from subsection 6.1 and denote the integers localized at p by $\mathbb{Z}_{(p)}$. We will work with the following inductive assumption.

Definition A.7.1. *Let $\mathcal{K}$ be an open G invariant subset of $\mathcal{S}(G)$ and X a smooth G manifold. We say that X is $\mathcal{K}$ adjusted if*

(1) *X^P is a $\mathbb{Z}_{(p)}$ homology sphere whenever p is a prime, $1 \neq P \in \mathcal{K} \cap \mathcal{P}(G)$, and P is of p power order,*
(2) *$\chi(X^H) = 2$ for all $H \in \mathcal{G}^1(G) \cap \mathcal{K}$, and*
(3) *X is a homotopy sphere if $\mathcal{K} = \mathcal{S}(G)$.*

This definition is related to our problem by the following observation.

Proposition A.7.2. *Let G be a finite group and Σ a homotopy sphere with smooth G action such that $\dim \Sigma^H$ is even for all $H \in \mathcal{S}(G)$. Then Σ is $\mathcal{S}(G)$ adjusted.*

Proof. The first condition of A.7.1 is a consequence of Smith theory [Br], and the last one is true by definition. We discuss the second one. Let x be a fixed point in Σ and $f : \Sigma \to S(\mathbb{R} \oplus T_x\Sigma)$ as in subsection A.4. The mapping cone, C_f, of f is contractible because f is a homotopy equivalence. The following argument is as in [O1]. Smith theory implies that C_f^P is a mod p disk whenever p is a prime and $P \in \mathcal{S}(G)$ is of p power order. In particular, $\chi(C_f^P) = 1$. Suppose $P \lhd H$ and H/P is cyclic, i.e., $H \in \mathcal{G}^1(G)$. Then $C_f^H = (C_f^P)^{H/P}$. The Lefschetz fixed point theorem implies that $\chi(C_f^H) = 1$, which means that $\chi(\Sigma^H) = S(\mathbb{R} \oplus T_x\Sigma)^H = 2$. □

A.8 Euler Characteristic Obstructions

The purpose of this subsection is to show

Proposition A.8.1. *Let X be a special G manifold with V structure and G an Oliver group. There exists an $(\mathcal{S}(G) \setminus \mathcal{P}(G))$ adjusted special G manifold X_1 with V structure such that $X^G = X_1^G$.*

Proof. Let $\delta \in \Omega(G)$ be an element as in Proposition A.1.2. Then

$$\begin{aligned}[(1+\delta) \ltimes X] - 2 &= (1+\delta)([X]-2) + (|\delta| - \delta)([X]-2) \\ &= (1+\delta)([X]-2) + 2\gamma\end{aligned}$$

Here $\gamma = \sum \gamma_H[G/H]$ where $\gamma_H = 0$ for all $H \notin \mathcal{G}^1(G)$. The first equality follows from A.6.3. Observe that $\chi_H(|\delta| - \delta) = 0$ if $H \notin \mathcal{G}^1(G)$ by our assumption on δ ($\delta_H \geq 0$ if $H \notin \mathcal{G}^1(G)$), and $(|\delta| - \delta)$ is divisible by 2. Hence $(|\delta| - \delta)([X]-2)$ may be written as 2γ for γ as above. We perform some surgeries on $(1+\delta) \ltimes X$, subtracting 2γ in $\Omega(G)$, to obtain X_1 such that $[X_1] = 2 + (1+\delta)([X]-2)$. Then, by assumption, $\operatorname{Res}_K^G(1+\delta) = 0$ and $\operatorname{Res}_K^G[X_1] = 2$ for all $K \in \mathcal{G}^1(G)$. This implies that $\chi(X_1^K) = 2$ for all $K \in \mathcal{G}^1(G)$, so X_1 is $(\mathcal{S}(G) \setminus \mathcal{P}(G))$ adjusted. □

A.9 More on the Operation $\ltimes$

We like to study the effect of the operation $\ltimes$ on equivariant surgery obstructions. Suppose X is a special G manifold with V structure (b, c) and X is $\mathcal{K}$ adjusted where $\mathcal{K} \supseteq \mathcal{S}(G) \setminus \mathcal{P}(G)$. Suppose also that V satisfies the strong gap condition in A.2.5. Let P be a maximal element in $\mathcal{S}(G) \setminus \mathcal{P}(G)$. Set $\dim X^P = 2k$. Suppose that X^P is $(k-1)$ connected, which can always be arranged with the help of equivariant surgery of isotropy type P. Abbreviate $H_k(X^P, \mathbb{Z})$ as H_k and $N_G P/P$ as W. We use Wall's geometric procedure [W1, Section 5] to define an intersection form λ and a self-intersection form μ on H_k. Specifically, set $\Lambda = \mathbb{Z}[W]$ and $Q = \Lambda/\{\nu - (-1)^k \overline{\nu} \mid \nu \in \Lambda\}$. Note that H_k is a module over Λ. Observe that the G action on V and X are orientation preserving as we assumed that V is the realification of a complex representation. We define the conjugation on Λ as follows. If $\nu = \sum a_g g$, then $\overline{\nu} = \sum a_g g^{-1}$. Then $\lambda : H_k \times H_k \to \Lambda$ and $\mu : H_k \to Q$. The forms are well defined due to the assumption that V, and with it X, satisfy the the strong gap condition. They also satisfy the properties spelled out in the conclusion of Theorem 5.2 of [W1]. Set

$$s(X, (b, c), P) = (H_k, \lambda, \mu). \tag{A.9.1}$$

Given a triple (H_k, λ, μ), we may apply restriction and induction (Res_C^W and Ind_C^W for $C \subseteq W$) to it. Specifically, $(\pm W/C)(H_k, \lambda, \mu)$ expresses that we restrict the action of W to an action of C and induce it back up to an action of W, and a negative sign indicates that we multiply the forms with -1.

Proposition A.9.2. *Let $\alpha = \sum \alpha_H (G/H)$ be a $\mathbb{Z}$ linear combination of homogeneous spaces with $\alpha_G = 0$. Otherwise, let the data be as described so far. Then*

(1) *$(1+\alpha) \ltimes (X, (b, c))$ is a $\mathcal{K}$ adjusted special G manifold with V structure,*
(2) *$((1+\alpha) \ltimes X)^P$ is $(k-1)$ connected, and*
(3) *$s((1+\alpha) \ltimes (X, (b, c)), P) = (1 + \alpha^P) s(X, (b, c), P)$.*

Proof. The claims follow easily from the explicit nature of the construction $\ltimes$. E. g., if X^H is a $\mathbb{Z}_{(p)}$ homology sphere, then in forming $(1+\alpha) \ltimes X$ we create several copies of X^H, and connect each of the newly created copies by a single zero dimensional surgery step to the copy of X^H in $1 \ltimes X$. Hence we again end up with a $\mathbb{Z}_{(p)}$ homology sphere. This, and the corresponding statement for Euler characteristics, verifies (1). To see (2) and (3) one observes that these zero dimensional surgeries only kill off classes in $\pi_0((1+\alpha) \odot X)$, and do not effect higher dimensional classes. For (3) we also remark that reversing G orientations (as dictated for negative α_H's) causes the intersection and self-intersection forms to be multiplied with (-1), and this is reflected on the algebraic side in the way we multiply (H_k, λ, μ) with $\pm W/C$. □

A.10 Wall Groups and Algebraic Induction

We recall some results from the literature (compare [DP3, p. 1384]). Let H be a finite group and $y(H, \mathbb{Z})$ and $GW(H, \mathbb{Z})$ as in [Dr2, p. 294]. Briefly, $y(H, \mathbb{Z})$ is obtained by applying the Grothendieck construction to the semiring of elements

(M, λ); M is a $\mathbb{Z}$ torsion free module over $\mathbb{Z}[H]$ and λ is an integral valued nonsingular symmetric bilinear form on M which satisfies $\lambda(hx, hy) = \lambda(x, y)$ for all $h \in H$ and x, $y \in M$. Then $GW(H, \mathbb{Z})$ is the residue class ring of $y(H, \mathbb{Z})$ with respect to the ideal generated by all split H-spaces in $y(H, \mathbb{Z})$. There are homomorphisms

$$\alpha : \Omega(G) \to y(H, \mathbb{Z}) \quad \text{and} \quad \beta : \Omega(G) \to GW(H, \mathbb{Z})$$

whose images are denoted by $\overline{\Omega}(H)$ and $\overline{\overline{\Omega}}(H)$. Define α as follows. Let S be a finite H set representing $[S]$ in $\Omega(H)$ and $\widetilde{S} = \mathbb{Z}[S]$ the free $\mathbb{Z}$ module with S as basis. This is a free $\mathbb{Z}$ module in an obvious way. It supports an integral valued symmetric bilinear form λ_S defined as follows. If s, $s' \in S$ then $\lambda_S(s, s') = 1$ if $s = s'$ and $\lambda_S(s, s') = 0$ if $s \neq s'$. The map β is the composition of α with the natural projection. Then $\alpha[S] = \overline{S}$ is the class of $(\widetilde{S}, \lambda_S)$ in $y(H, \mathbb{Z})$ and $\beta[S] = \overline{\overline{S}}$ is the class in $GW(H, \mathbb{Z})$. As α and β are multiplicative, $\overline{\Omega}(H)$ and $\overline{\overline{\Omega}}(H)$ are rings.

The induction homomorphism in the Burnside ring which arises from sending a finite H set S to the G set $G \times_H S$ induces induction homomorphisms $\mathrm{Ind}_H^G : \overline{\Omega}(H) \to \overline{\Omega}(G)$ and $\mathrm{Ind}_H^G : \overline{\overline{\Omega}}(H) \to \overline{\overline{\Omega}}(G)$ when $H \subset G$. Suppose we have modules $M(H)$ over rings $R(H)$ for all subgroups H of G, and induction and restriction homomorphisms Ind_H^G and Res_H^G for each of them; they have the obvious range and target. We say that $\{R(H),\ M(H) \mid H \subset G\}$ satisfies Frobenius reciprocity if for all $H \subset G$, $u \in R(G)$, $u' \in R(H)$, $v \in M(G)$, and $v' \in M(H)$ (see [Sw, p. 17])

$$\begin{aligned}
\mathrm{Res}_H^G(u \cdot v) &= \mathrm{Res}_H^G u \cdot \mathrm{Res}_H^G v \\
u \cdot \mathrm{Ind}_H^G v' &= \mathrm{Ind}_H^G(\mathrm{Res}_H^G u \cdot v') \\
\mathrm{Ind}_H^G u' \cdot v &= \mathrm{Ind}_H^G(u' \cdot \mathrm{Res}_H^G v).
\end{aligned}$$

A group is said to be hyperelementary if it has a cyclic normal subgroup of prime power index. Let $\mathcal{G}_1(G)$ denote the set of hyperelementary subgroups of G. In [DP3, Lemma 4.5 (vi)] we showed

Lemma X.10.1. *Let G be a finite group. Then*

$$1 \in \mathrm{im}\left(\bigoplus_{H \in \mathcal{G}_1(G)} \overline{\overline{\Omega}}(H) \xrightarrow{\mathrm{Ind}} \overline{\overline{\Omega}}(G) \right).$$

The map Ind is a collection of induction maps Ind_H^G for the indicated subgroup of G.

We need to define a subgroup $B_0(G)$ of the reduced projective class group $\widetilde{K}_0(\mathbb{Z}[G])$, where $\widetilde{K}_0(\mathbb{Z}[G])$ is regarded as the group $(\oplus)$ of finitely generated projective $\mathbb{Z}[G]$ modules modulo free modules [Sw]. A G resolution is a finite G CW complex A with a base point $q \in A^G$, which is k dimensional and $(k-1)$ connected for some k, and for which $\widetilde{H}_k(A)$ is $\mathbb{Z}[G]$ projective. Set

$$\gamma_G(A) = (-1)^k[\widetilde{H}_k(A)] \in \widetilde{K}_0(\mathbb{Z}[G]).$$

There exists a finite G CW complex A^* which contains A, obtained from A by attaching free G cells $G \times D^i$, for which $\chi(A^*) = 1$. Set $\{A\} = [A^*] - 1 \in \Omega(G)$. Then, following [O2] or [OP], we set

$$B_0(G) = \{\gamma_G(A) \mid A \text{ is a } G \text{ resolution and } \{A\} = 0\}.$$

Let P be a subgroup of G of p-power order, p prime. To P we associate two rings and a Wall group. We need the groups only in the even dimensional setting, so $d = 2k$ will always be even.

(A.10.2)
$$A_P = \begin{cases} \mathbb{Z} & \text{if } P = 1 \\ \mathbb{Z}_{(p)} & \text{if } P \neq 1 \end{cases}$$
$$\Lambda_P = A_P[NP/P]$$

(A.10.3)
$$J_G(P,d) = \begin{cases} L_d^{B_0(G)}(\mathbb{Z}[G], 1) & \text{if } P = 1 \\ L_d(A_P[NP/P], 1) & \text{if } P \neq 1 \end{cases}$$

Essentially, the definition of $J_G(P,d)$ is as in [W, Section 5]. The elements are represented by triples (H, λ, μ) where H is a Λ_P module and $\lambda : H \times H \to \Lambda_P$ is a $(-1)^k$ hermitian form with associated quadratic form $\mu : H \to Q_P = \Lambda_P/\{\nu - (-1)^k\overline{\nu} \mid \nu \in \Lambda_P\}$. As above, Λ_P comes with a natural antiautomorphism $^-$. The underlying module H is assumed to be (stably) free over Λ_P if $P \neq 1$. If $P = 1$, H is assumed to be a finitely generated, projective $\Lambda_1 = \mathbb{Z}[G]$ module whose class in $\widetilde{K}_0(\mathbb{Z}[G])$ lies in $B_0(G)$. The class of (H, λ, μ) represents zero in $J_G(P,d)$ if it is stable isomorphic to the hyperbolic form on a module M where M is free over Λ_P if $P \neq 1$. If $P = 1$, M is assumed to be a finitely generated, projective $\mathbb{Z}[G]$ module whose class in $\widetilde{K}_0(\mathbb{Z}[G])$ lies in $B_0(G)$. To be more precise, particularly concerning the relation between λ and μ, one can use the setup in [W2, Section 1.1] or [Ra, Section 1] to define $J_G(P,d)$, compare [DP3, Section 4]. It is important to observe that

Proposition A.10.4. *The collections*

$$\{\overline{\overline{\Omega}}(H),\ L_d^{B_0(H)}(\mathbb{Z}[H], 1) \mid H \subset G\} \quad \textit{and} \quad \{\overline{\overline{\Omega}}(C),\ L_d(\mathbb{Z}_{(p)}[C], 1) \mid C \subset NP/P\}$$

satisfy Frobenius reciprocity.

Proof. This is a routine exercise. One needs to retrace the definitions. Still, it is useful to remember that $\overline{\overline{\Omega}}(H)$ is a subring of $GW(H, \mathbb{Z})$ with induction and restriction maps obtained from the corresponding maps for $\Omega(H)$, and these are the same as the restrictions of the induction and restriction maps for the Grothendieck-Witt rings. For the first collection we also note that there are induction and restriction maps for $B_0(G)$ and that $\{\overline{\overline{\Omega}}(H),\ B_0(H) \mid H \subset G\}$ satisfies Frobenius reciprocity, see also [DP3, Lemma 4.5 (i)]. □

To use the results of this subsection, we need furthermore

Lemma A.10.5. *Let G be an Oliver group, P a subgroup of G of prime power order, and β as above. There exists an element $\delta = \sum \delta_H[G/H]$ in $\Omega(G)$ such that $\delta_G = 0$ and $\beta(1 + \delta^P) = 0$ in $GW(W, \mathbb{Z})$.*

Proof. If $P = 1$, then we only need to note that G is not hyperelementary, and the claim follows from Proposition A.10.1.

Now suppose that $P \neq 1$. Denote $N_G P/P$ by W and the projection $NP \to W$ by α. Let p be the prime such that P is of p power order. Lemma A.10.1 says that there exists an element $\Delta = \sum \Delta_K[W/K] \in \Omega(W)$, where K varies over the hyperelementary subgroups of W, such that $\beta(\Delta) = -1 \in GW(W,\mathbb{Z})$. Set $\widetilde{K} := \alpha^{-1}(K)$ and $\delta = \sum \Delta_K[G/\widetilde{K}]$. Then, as elements in $\Omega(W)$, $\delta^P = \Delta$, and $\beta(1+\delta^P) = 0$ in $GW(W,\mathbb{Z})$. We need to show that the coefficient δ_G of $[G/G]$ in δ is zero. To see this, consider a hyperelementary subgroup K of W and a cyclic normal subgroup C of K of q power index, q prime. Then $\widetilde{K}$ is in $\mathcal{G}_p^q(N_G P)$ ($\mathcal{G}_p^q$ is defined before Proposition A.1.2). In particular, $\widetilde{K} \neq G$ because G is assumed to be an Oliver group. This means that δ^G is empty and $\delta_G = 0$. □

A.11 Geometric Surgery obstructions

The setting in this subsection is as follows. Let $\mathcal{K}$ be an open G invariant subset of $\mathcal{S}(G)$ such that $\mathcal{S}(G) \setminus \mathcal{K}$ contains only groups of prime power order and P a maximal element in $\mathcal{S}(G)\setminus\mathcal{K}$ (i. e., if $P \subset H$, $H \neq P$, then $H \in \mathcal{K}$). If $P \neq 1$, then let p be the prime such that P is of p power order. Let $\mathcal{K}' = \mathcal{K} \cup \{gPg^{-1} \mid g \in G\}$. Suppose that X is a special G manifold with V structure which is $\mathcal{K}$ adjusted, $\dim X^P = 2k$, and V satisfies the strong gap condition. Assume that X^P is $(k-1)$ connected, which is always possible after some surgeries of isotropy type P. Let $s(X,(b,c),P)$ be as in A.9.1, and set

(A.11.1)
$$\sigma((X,(b,c),P) = \begin{cases} [s(X,(b,c),P)_{(p)}] \in L_{2k}(\mathbb{Z}_{(p)}[NP/P],1) & \text{if } P = 1 \\ [s(X,(b,c),P)] \in L_{2k}^{B_0(G)}(\mathbb{Z}[G],1) & \text{if } P \neq 1 \end{cases}$$

For the first part of the definition we note that $\chi(X^H) = 2$ whenever there is a short exact sequence $1 \to P \to H \to C \to 1$ where C is cyclic, and this implies that $H_k(X^P,\mathbb{Z}_{(p)})$ is a free $\mathbb{Z}_{(p)}[NP/P]$ module [OP, Lemma 2.5]. The subscript (p) indicates that the data $s(X,(b,c),P)$ is used with $\mathbb{Z}_{(p)}$ coefficients, and [ ] denotes the class in the indicated Wall group. If $P = 1$, then it follows as in [DP3, 5.1 and p. 1390] that the class of $H_k(X,\mathbb{Z})$ in $\widetilde{K}_0(\mathbb{Z}[G])$ lies in $B_0(G)$. In either case $\sigma((X,(b,c),P)$ is an element in the indicated Wall group.

Remark. As pointed out in [Mo3, p. 466] and clarified in [DS, p. 31] (see also [D2]), $\sigma((X,(b,c),P)$ may not be well defined under borderline gap conditions unless the group is of odd order, the prevailing assumption in [DP3]. Our gap condition (SGC, see A.2.5) is stronger than the one in [DP3], and all of the results on $\sigma((X,(b,c),P)$ quoted from this reference hold in our context, also for even order groups.

Proposition A.11.2 ([DP3, Lemma 5.3]). *Let all data be as above. Suppose $\sigma((X,(b,c),P) = 0$, then, as a special G manifold with V structure, X is equivariantly cobordant to a special G manifold X_1 which is $\mathcal{K}'$ adjusted. The cobordism is relative to the K fixed point set for all $K \in \mathcal{K}$.*

Proposition A.11.3. *Let X be special G manifold with V structure which is $\mathcal{K}$ adjusted, and let $\mathcal{K}$, P and $\mathcal{K}'$ be as above. There exists a special G manifold X_1 with V structure which is $\mathcal{K}'$ adjusted, and $X_1^G = X^G$.*

Proof. After some equivariant surgeries of isotropy type P, we may suppose the X^P is $(k-1)$ connected. Then $\sigma(X,(b,c),P)$ is defined. Let δ be as in Lemma

A.10.5. Then $\sigma((1+\delta)\times(X,(b,c),P))=0$, see Proposition A.9.2. Next we apply Proposition A.11.2 to $(1+\delta)\times(X,(b,c),P)$ to find a special G manifold X_1 with V structure which is $\mathcal{K}'$ adjusted, and $X_1^G = X^G$. □

A.12 Proof of the Theorem

Finally, we put the tools developed so far together to prove the theorem of Laitinen and Moriomoto under our special assumptions.

Theorem A.12.1. *Suppose G is an Oliver group and G has an admissable representation which satisfies the strong gap condition in A.2.5. Then G acts smoothly on the standard sphere in some dimension with exactly one fixed point.*

The dimension of the sphere may be taken as the n in A.2.5, i. e., as the dimension of the admissable representation used in the construction. As verified in A.2.5, the assumptions hold if G is an Oliver group which is 2-perfect (e.g., if G is of odd order or simple) or nilpotent (e.g., if G is abelian). First of all, we show that it suffices to construct an action on a homotopy sphere.

Lemma A.12.2. *Let G be a finite group not of prime power order, and Σ a homotopy sphere with a smooth G action. Assume that G_p is an isotropy group and Σ^{G_p} is connected and of dimension at least 2 for each Sylow p-subgroup. Then G acts on the standard sphere S of the same dimension as Σ and $S^H = \Sigma^H$ for all subgroups H of G not of prime power order.*

Proof. Let Λ be the set of all primes dividing the order of G and $a_p := [G : G_p]$ for $p \in \Lambda$. Note that $\{a_p\}_{p\in\Lambda}$ is a set of relatively prime numbers. Denote the order of Σ in the group of homotopy spheres by m. There is a set of natural numbers $\{\alpha_p\}_{p\in\Lambda}$ such that $\sum \alpha_p a_p \equiv -1 \pmod{m}$. Set

$$Z = \Sigma \sqcup \bigsqcup_{p\in\Lambda} \alpha_p (G \times_H \Sigma).$$

Then $Z^H = \Sigma^H$ if H is not of prime power order, Z is equivariantly cobordant to a homotopy sphere Σ_1, and the cobordism is relative to the H fixed point set for all subgroups H of G not of prime power order. Specifically, Σ_1 is constructed from Z by zero dimensional surgeries on $\bigsqcup_{p\in\Lambda} \alpha_p(G\times_H S^0)$, where $\bigsqcup \alpha_p(G\times_H y_+)$ is equivariantly embedded in Σ and $\bigsqcup \alpha_p(G\times_H y_-)$ in $\bigsqcup \alpha_p(G\times_H \Sigma)$. Here $y_\pm$ denote the poles of S^0. The cobordism is the trace of the surgeries. By construction, Σ_1 represents m times the class of Σ in the group of homotopy spheres, and this means that Σ_1 is diffeomorphic to the standard sphere. □

Proof of Theorem. Let G be an Oliver group and V an admissable representation of G, see A.2.5. Theorem A.5.1 provides us with a closed smooth G manifold X with V structure such that the action has exactly one fixed point and X^H is connected whenever it is of positive dimension. We introduce pairs of cancelling points for all isotropy groups H, $H \neq G$, for which $\dim V^H = 0$, see subsection A.3, and obtain a special G manifold X_1 with V structure and exactly one fixed point. Apply Proposition A.8.1 to obtain a special G manifold X_2 with V structure and exactly one fixed point which is $(\mathcal{S}(G)\setminus\mathcal{P}(G))$ adjusted.

Apply induction over the partial ordering on the G invariant open subsets of $\mathcal{S}(G)$ containing $\mathcal{S}(G) \setminus \mathcal{P}(G)$, and use Proposition A.11.3 to carry out the inductive step. In each step we enlarge the set $\mathcal{K}$ of subgroups for which the manifold is adjusted. As a result we obtain a homotopy sphere with smooth G action and exactly one fixed point. Finally, apply Lemma A.12.2 to the action on the homotopy sphere to get the smooth one fixed point action on a standard sphere. □

REFERENCES

[BM] A. Bak and M. Morimoto, *Equivariant Surgery on Compact Manifolds with Half Dimensional Singular Sets*, preprint 1991.

[Br] G. E. Bredon, *Introduction to Compact Transformation Groups*, Academic Press, New York and London, 1972.

[BKS] N. P. Buchdale, S. Kwasik, R. Schultz, *One Fixed Point Actions on Low-Dimensional Spheres*, Invent. Math. **102** (1990), 633–662.

[DeM] S. DeMichelis, *The Fixed Point Set of a Finite Group Action on a Homology Four Sphere*, L'Enseign. Math. **35** (1989), 107–116.

[tD1] T. tom Dieck, *The Burnside Ring of a Compact Lie Group I*, Math. Ann. **215** (1975), 235–250.

[tD2] T. tom Dieck, *Transformation Groups*, de Gruyter Studies in Mathematics 8, Walter de Gruyter, Berlin, New York, 1987.

[D1] K. H. Dovermann, *Addition of Equivariant Surgery Obstructions*, Ph.D. Thesis, Rutgers University, 1978 (Available from University Microfilms, Ann Arbor, Mich.: Order Number DEL79-10380. Summarized in Dissertation Abstracts International **39** (1978/1979), B5406).

[D2] K. H. Dovermann, *Almost Isovariant Normal Maps*, Amer. J. Math. **111** (1989), 851–904.

[DP1] K. H. Dovermann and T. Petrie, *G Surgery II*, Memoirs of the Amer. Math. Soc. **30** no 260 (1982).

[DP2] K. H. Dovermann and T. Petrie, *Smith Equivalence of Representations for Odd Order Cyclic Groups*, Topology **24** (1985), 283–305.

[DP3] K. H. Dovermann and T. Petrie, *An Induction Theorem for Equivariant Surgery (G-Surgery III)*, Amer. J. Math. **105** (1983), 1369–1403.

[DR] K. H. Dovermann and M. Rothenberg, *Equivariant Surgery and Classification of Finite Group Actions on Manifolds*, Memoirs of the Amer. Math. Soc. **71** no 379 (1988).

[DS] K. H. Dovermann and R. Schultz, *Equivariant Surgery Theories and their Periodicity Properties*, Lecture Notes in Math. Vol. 1443, Springer Verlag, Berlin, Heidelberg, New York, London, Paris, Tokyo, Hong Kong, Barcelona, 1990.

[Dr1] A. Dress, *Contributions to the Theory of Induced Representations*, Algebraic K-Theory II, Lecture Notes in Math., vol. 342, Springer Verlag, Berlin, Heidelberg, New York, 1973, pp. 183–240.

[Dr2] A. Dress, *Induction and Structure Theorems for Orthogonal Representations of Finite Groups*, Ann. Math. **102** (1975), 291–325.

[Fu] M. Furuta, *A Remark on a Fixed Point of Finite Group Actions on S^4*, Topology **28** (1989), 35–38.

[L] E. Laitinen, *Unstable Homotopy Theory of Homotopy Representations*, Transformation Groups Poznań 1985 (S. Jackowski and K. Pawalowski, eds.), Lecture Notes in Math., vol. 1217, Springer Verlag, Berlin, Heidelberg, New York, London, Paris, Tokyo, 1986, pp. 210–248.

[LM] E. Laitinen and M. Morimoto, *Finite Groups with Smooth One Fixed Point Actions on Spheres*, preprint 1993.

[LMP] E. Laitinen, M. Morimoto, and K. Pawalowski, *Smooth Actions of Finite Nonsolvable Groups on Spheres*, preprint 1992.

[MS] D. Montgomery and H. Samelson, *Fiberings with Singularities*, Duke Math. J. **13** (1946), 51–56.

[Mo1] M. Morimoto, *One Fixed Point Actions on Sphere*, Proc. Japan Acad. Ser. A **63** (1987), 95–97.

[Mo2] M. Morimoto, *S^4 Does Not Have One Fixed Point Actions*, Osaka J. Math. **25** (1988), 575–580.

[Mo3] M. Morimoto, *Bak Groups and Equivariant Surgery*, K-Theory **2** (1989), 456–483.

[O1] R. Oliver, *Fixed Point Sets of Group Actions on Finite Acyclic Complexes*, Comment. Math. Helv. **50** (1975), 155–177.

[O2] R. Oliver, *Group Actions on Disks, Integral Permutation Representations, and the Burnside Ring*, Proc. of Symposia in Pure Math., vol. XXXII, Part 1, Amer. Math. Soc., 1978, pp. 339–346.

[OP] R. Oliver and T. Petrie, *G CW Surgery and $K_0(\mathbb{Z}[G])$*, Math. Zeitschrift **179** (1982), 11–42.

[Pa] K. Pawalowski, *Fixed Point Sets of Smooth Actions on Disks and Euclidean Spaces, A Survey*, Geometric and Algebraic Topology, Banach Center Publications, vol. 18, Polish Scientific Publishers, Warsawa, 1986, pp. 165–180.

[P1] T. Petrie, *One Fixed Point Actions on Spheres, I*, Adv. Math. (1982), 3–14.

[P2] T. Petrie, *One Fixed Point Actions on Spheres, II*, Adv. Math. (1982), 15–70.

[P3] T. Petrie, *Pseudoequivalences of G-manifolds*, Proc. of Symposia in Pure Math., vol. XXXII, Part 1, Amer. Math. Soc., 1978, pp. 169–186.

[P4] T. Petrie, *G-transversality and Applications*, Proc. of Symposia in Pure Math., vol. XXXII, Part 1, Amer. Math. Soc., 1978, pp. 186–210.

[PR] T. Petrie and J. Randall, *Transformation Groups on Manifolds*, Dekker Series in Pure and Applied Mathematics, Vol. 82, Marcel Dekker, New York, 1984.

[Ra] A. Ranicki, *Algebraic L Theory I: Foundations*, Proc. London Math. Soc. (3) **27** (1973), 101–125.

[Se] G. Segal, *Equivariant Stable Homotopy Theory*, Proc. International Congress of Math. (Nice 1970), Gauthier-Villars, Paris, 1971, pp. 59–63.

[Sw] R. Swan, *K-Theory of Finite Groups and Orders*, Lecture Notes in Math. Vol. 149, Springer Verlag, Berlin, Heidelberg, New York, 1970.

[W1] C. T. C. Wall, *Surgery on Compact Manifolds*, London Math. Soc. Monographs Vol. 1, Academic Press, London and New York, 1970.

[W2] C. T. C. Wall, *Classification of Hermitian Forms, VI Group Rings*, Ann. of Math. **103** (1976), 1–80.

Department of Mathematics, University of Hawaii, Honolulu, HI 96822, USA
E-mail address: dover@ uhunix.bitnet

Contemporary Mathematics
Volume **182**, 1995

A Numerical Approach to Defining a Legendre Map in Genus 2

R. SILHOL

ABSTRACT. In this paper we describe a test version of an algorithm to compute an equation for a genus 2 curve admitting a given period matrix.

1. Introduction

Let C be a smooth and projective algebraic curve of genus g. Let $(a_1, \ldots, a_g, b_1, \ldots, b_g)$ be cycles defining a standard basis of $\mathrm{H}_1(C, \mathbb{Z})$ (i.e, such that the matrix of the intersection form is in standard form) and let $(\omega_1, \ldots, \omega_g)$ be a basis of the space of holomorphic differential 1-forms on C. Let

$$X = \left(\int_{a_j} \omega_i \right) \quad \text{and} \quad Y = \left(\int_{b_j} \omega_i \right) .$$

Then a period matrix of the curve C is,

$$Z = X^{-1} Y .$$

The importance of such a matrix is that, it completely determines the curve C up to isomorphy.

Strangely, except in genus 1 or for very special curves, an explicit relation between a curve and its period matrix is unknown. For elliptic curves we have the Weierstrass map defined as follows. Let $\tau \in \mathbb{C}$, with $\Im m(\tau) > 0$, and let

$$g_k(\tau) = \sum_{(n,m) \in \mathbb{Z}^2 \setminus (0,0)} (n + m\tau)^{-2k} .$$

Then, τ is a period for the plane curve defined by,

$$y^2 = 4x^3 - g_2(\tau)x - g_3(\tau) .$$

1991 *Mathematics Subject Classification.* Primary 32G20; Secondary 14H40.

Supported in part by EC HCM Plan No. CHRX-CT93-0408

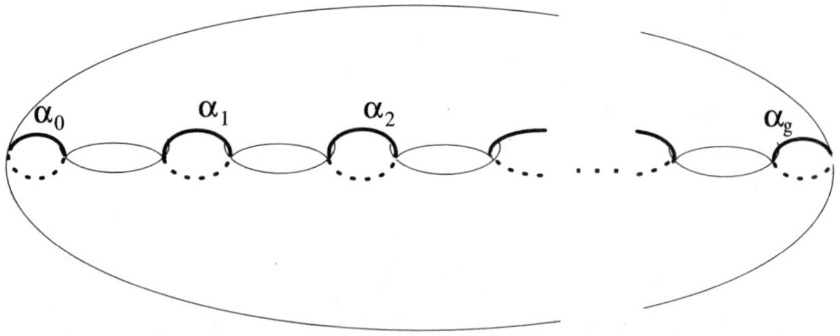

FIGURE 1.

Closely related to the Weierstrass map is the Legendre map $\lambda(\tau)$ (see for example A. Robert [4]) which associates to τ the curve,

$$y^2 = x(x-1)(x-\lambda(\tau)) ,$$

for which τ is a period.

The aim of this paper is to give a brief, and partially informal, description of a method to recover from a 2×2 period matrix Z the equation of a genus 2 curve having Z as period matrix, or to be more precise, a numerical approximation of the roots of a polynomial $P(x)$ such that the curve defined by $y^2 = P(x)$ has Z as period matrix. Actually in this introductory paper we will limit ourselves to one case, the case of real curves with 3 real components. A more general treatment, with full proofs, will appear in [2].

2. Computing the period matrix of a hyperelliptic curve

The core of the method relies on a method to compute a numerical approximation for a period matrix of an algebraic curve[1] that we describe in this section (see [7]).

A hyperelliptic curve C is a ramified double covering of $\mathbb{P}^1(\mathbb{C})$ and can always be represented by an equation of the form, $y^2 = P(x)$, where P is a polynomial of degree $2g+2$ with distinct roots $(x_1, \ldots, x_{2g+2})$. One can order these roots in such a way that the broken line joining x_1 to x_2, x_2 to x_3, ... , x_{2g+1} to x_{2g+2} does not self intersect.

Call α_i (resp. β_i) the cycle (unoriented for the moment) obtained by taking the pullback in C of the line segment joining x_{2i+1} to x_{2i+2} (resp. x_{2i} to x_{2i+1}).

Now C is a double covering of $\mathbb{P}^1(\mathbb{C})$ with the x_i's as branch points (the covering is given by the map $(x,y) \mapsto x$). From this it is relatively straightforward to see that we can interpret the α_i's and β_i's as indicated in figures 1 and 2.

To orient the cycles is theoretically easy one only needs to note that at the points $(x_i, 0)$, y is a local coordinate on the curve. In other words the orientation is given by $(\Re e(y), \Im m(y))$. In practice it is a problem of selecting half branches

[1]Developed with partial support from the EC Science Plan SC1*-CT91-0716.

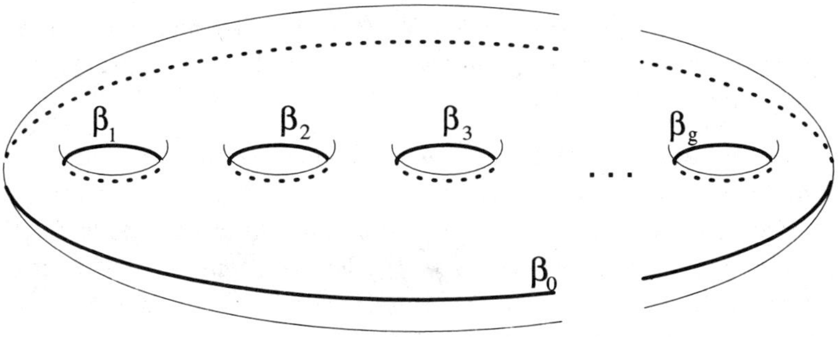

FIGURE 2.

or in other words determinations of the square root of $P(x)$. This is a mild technical problem and a practical solution is described in [**7**]. Once an orientation is chosen it is easy to recover from the α_i's and β_i's a standard basis of the homology of C.

Now it is well known that,

$$\frac{x^r\,dx}{y} \quad \text{where } 0 \le r < g\ ,$$

form a basis of the space of holomorphic differentials on C (see for example Mumford [**3**]) so finally the problem of computing a period matrix of C is reduced to a problem of numerical integration which is easy to solve.

3. Period matrices of genus 2 curves

Period matrices have special properties in particular they are symmetric (by Riemann's equality) and their imaginary part is positive definite (by Riemann's inequality). Unfortunately this is not enough in general to characterize the period matrix of a curve and the problem of characterizing such period matrices (Shottky's problem) is a very difficult one. In genus 2 (and also in genus 3) however the problem is easier since period matrices are dense in the set $\mathcal{H}_2$ of complex 2 by 2 symmetric matrices with positive definite imaginary part.

More precisely the situation is as follows. Let $\mathbf{Sp}_4(\mathbb{Z})$ be the symplectic group and let $N = \left(\begin{smallmatrix} a & b \\ c & d \end{smallmatrix}\right)$, where a, b, c and d are 2×2 integer matrices. $\mathbf{Sp}_4(\mathbb{Z})$ operates on $\mathcal{H}_2$ by $N(Z) = (aZ+b)(cZ+d)^{-1}$. With these notations we have, the well known,

LEMMA 3.1. *$Z \in \mathcal{H}_2$ is the period matrix of a smooth algebraic curve, if and only if, Z is not of the form $N(Z')$, where $N \in \mathbf{Sp}_4(\mathbb{Z})$ and*

$$Z' = \begin{pmatrix} x+iy & 0 \\ 0 & x'+iy' \end{pmatrix} .$$

REMARK 3.2. If Z is of the form $N(Z')$, Z' as in (3.1) then, Z corresponds to the singular curve obtained by gluing at one point the two elliptic curves with respective periods $x + iy$ and $x' + iy'$.

This still does not completely solve the problem since we should find all orbits of the subspace of such matrices under the action of $\mathbf{Sp}_4(\mathbb{Z})$. But there is a case where it does, namely if the curve is real, because in this case we have an efficient way of normalizing period matrices (see [**6**]). For real genus 2 curves with 3 real components (which is the only case we will consider in this introductory paper), the description is quite easy. One can always normalize the period matrix Z in such a way that, the real part is 0 and the imaginary part satisfies,

$$0 < y_1 \leq y_2,\ 0 \leq 2y_{12} \leq y_1, \quad \text{where } Z = \begin{pmatrix} iy_1 & iy_{12} \\ iy_{12} & iy_2 \end{pmatrix} . \tag{3.3}$$

In the subspace defined by (3.3) the matrices which are not period matrices of smooth curves are exactly those with $y_{12} = 0$. Note also that these matrices correspond each to one and only one real isomorphism class of real curves.

For practical reasons we will work in a somewhat larger space, defined by,

$$0 < y_1 \leq y_2,\ 0 < y_{12} < y_1 . \tag{3.4}$$

It is obtained by gluing the space defined by (3.3) and its images under,

$$\begin{pmatrix} 0 & 1 & 0 & 0 \\ -1 & 1 & 0 & 0 \\ 0 & 0 & 1 & 1 \\ 0 & 0 & -1 & 0 \end{pmatrix} \quad \text{and} \quad \begin{pmatrix} 1 & 0 & 0 & 0 \\ 1 & -1 & 0 & 0 \\ 0 & 0 & 1 & 1 \\ 0 & 0 & 0 & -1 \end{pmatrix} \tag{3.5}$$

Note that both matrices in (3.5) transform matrices with real part 0 and $y_{12} = 0$ into matrices with real part 0 and $y_{12} = y_1$. In other words all matrices in the space defined by (3.4) are period matrices of smooth curves. The reason why we consider this space rather than the space defined by (3.3), is that the matrices computed by the program mentioned above, lie in this space.

The space defined by (3.5) contains some remarkable subspaces, in particular the ones defined by $y_1 = y_2$ and $y_1 = y_2 = 2y_{12}$ (there are others, but we will only consider these two). Note that the first is point-wise fixed by,

$$\begin{pmatrix} 0 & 1 & 0 & 0 \\ 1 & 0 & 0 & 0 \\ 0 & 0 & 0 & 1 \\ 0 & 0 & 1 & 0 \end{pmatrix}$$

and the second is in addition fixed by the matrices defined in (3.5).

Since by Torelli's theorem a curve admits a non-trivial automorphism if and only if its period matrix is fixed by an element of $\mathbf{Sp}_{2g}(\mathbb{Z})$ different from ± 1 (-1 corresponding to the hyperelliptic involution, when it exists) we see that the curves with period matrix satisfying $y_1 = y_2$ admit an involution (different from the hyperelliptic involution) and the ones with $y_1 = y_2 = 2y_{12}$ admit a subgroup of automorphisms isomorphic to the dihedral group D_3. The important point is that one can find explicit representatives for these curves.

LEMMA 3.6. *Let C be a real genus 2 curve with real automorphism group isomorphic to $D_3 \times \mathbb{Z}/2$, then C is real isomorphic to a curve with equation,*

$$y^2 = (x^2 - a^2)(x^2 - b^2)(x^2 - c^2) ,$$

where,

$$1 < a < 3, \quad b = \frac{3+a}{a-1} \quad \text{and} \quad c = \frac{3-a}{a+1} .$$

Moreover such a real representative is unique. On the other hand, from the complex point of view, curves corresponding to a and $3/a$ are complex isomorphic.

PROOF. The curves described in (3.6) have a real automorphism of order 3, defined by,

$$(x, y) \longmapsto \left(\frac{x+3}{1-x}, \frac{32}{(1-x)^3} \cdot y\right) .$$

By [1] we know that genus 2 curves admitting an automorphism of order 3, have a real automorphism group isomorphic to $D_3 \times \mathbb{Z}/2$.

By the remarks made above, the curves described in (3.6) have a period matrix satisfying $y_1 = y_2 = 2y_{12}$. When a tends to 1 the components of the curve degenerate to points. By [5] this means that the coefficients of the period matrix tend to infinity. On the other hand when a tends to 3 the curve acquires 3 ordinary quadratic singularities with real tangents. Again by [5] this means that the coefficients of the period matrix tend to 0. This means that when a varies from 1 to 3 the period matrices of the curves describe the full line defined by $y_1 = y_2 = 2y_{12}$. Since this is a subset of the moduli space, in particular each such matrix corresponds to one and only one real isomorphism class of curves, we have all but the last claim. The last follows from the fact that Z and $-Z^{-1}$ correspond to complex isomorphic curves (but in general not real isomorphic) and an easy computation proving that for curves corresponding to a and $3/a$ the map, $x \mapsto 3/x$, induces a complex isomorphism.

LEMMA 3.7. *A real genus 2 curve with 3 real components and with a real involution distinct from the hyperelliptic involution is real isomorphic to one with equation,*

$$y^2 = (x^2 - 1)(x^2 - a^2)(x^2 - b^2) .$$

PROOF. First recall that for hyperelliptic curves defined by an equation of the form $y^2 = P(x)$ a non trivial automorphism is always induced by a map $x \mapsto \varphi(x)$, where $\varphi \in \mathrm{PGL}_1$ leaves the roots of P globally fixed. On the other hand involutions of PGL_1 are conjugate to $\left(\begin{smallmatrix} -1 & 0 \\ 0 & 1 \end{smallmatrix}\right)$. This proves the claim.

4. Computing an equation of the curve corresponding to a given period matrix

The complete method to compute equations of curves of genus 2 admitting a given symmetric, with positive imaginary part, 2×2 complex matrix is being written by Y. Chekouri [**2**]. Here we will only describe the special case of real curves with 3 real components or in other words matrices with real part equal to zero and imaginary part satisfying (3.4).

We first note that by (3.6) the one dimensional subspace defined by the conditions $y_1 = y_2 = 2y_{12}$ corresponds to the one parameter family described in (3.6) and that, for the reasons mentioned above, increasing a will decrease y_1. In this case, using the program for computing period matrices described in **1**, the computation is easy and a variety of methods will work. One can also somewhat improve the program by noting that the matrix,

$$\begin{pmatrix} \frac{2i}{\sqrt{3}} & \frac{i}{\sqrt{3}} \\ \frac{i}{\sqrt{3}} & \frac{2i}{\sqrt{3}} \end{pmatrix}$$

corresponds to $a = \sqrt{3}$ and that when y_1 is larger (resp. smaller) than $\frac{2i}{\sqrt{3}}$, a is smaller (resp. larger) than $\sqrt{3}$.

In the moduli space of real genus 2 curves, with 3 real components (see (3.3)), those admitting a non trivial automorphism correspond to $y_1 = y_2$ (and $2y_{12} \leq y_2$) or $y_1 = 2y_{12}$ (and $y_1 \leq y_2$). The first matrix of (3.5) transforms the matrices satisfying $y_1 = 2y_{12}$ and $y_1 \leq y_2$ into matrices satisfying $y_1 = y_2$ and $y_2 \leq 2y_{12}$. By (3.7) the full subspace of matrices satisfying $y_1 = y_2$ and $0 < y_{12} < y_1$ corresponds to the curves defined by an equation of the form $y^2 = (x^2 - 1)(x^2 - a^2)(x^2 - b^2)$. Moreover transforming by $x \mapsto x/a$ or $x \mapsto x/b$ we may always assume that $0 < a < 1$ and $b > 1$. Also for such curves the program mentioned in **2** will return a period matrix satisfying the above conditions. To understand how the y's vary in terms of the coefficients of the equation we again study the possible degenerations.

Assume first that b is fixed. If a goes to zero the curve acquires an ordinary quadratic singularity with real tangents. By [**5**] this implies that y_{12} will tend to y_1 and y_1 will tend to the period of the elliptic curve defined by $y^2 = (x^2 - 1)(x^2 - b^2)$. On the other hand if a tends to 1 the curve will degenerate to one with 2 isolated singular points (quadratic singularities with complex conjugate tangents). Again by [**5**] this means that $y_1 = y_2$ will tend to infinity while y_{12} will tend to a finite limit.

In other words in the space $y_1 = y_2$ the curve described by the period matrices of $y^2 = (x^2 - a^2)(x^2 - 1)(x^2 - b^2)$, b fixed, has the general aspect of a branch of a hyperbola. In fact experimental data show that the product $y_1 \cdot y_{12}$ varies very little, at least if a is not to close to 1. For example for $b = 3$ and $a = 0.3$, 0.4, 0.5, 0.6 we find respectively $y_1 y_{12} = 0.543807$, 0.537651, 0.535731 and 0.538939.

This heuristic remark gives us a way to get a first estimate on the size of b. From this, finding an approximation of a and b is just a question of adjusting coefficients noting, for reasons explained above that if a increases, y_1 will increase and y_{12} will decrease and for similar reasons if b increases, both y_1 and y_{12} will increase.

Now let C be a general real genus 2 curve with 3 real components. For C we can always find an equation of the form

$$y^2 = (x+a)(x+1)(x+b)(x-c)(x-1)(x-d) \ ,$$

a, b, c and d positive, a and d greater than 1, b and c less than 1. To C we can associate 2 curves C' and C'' defined respectively by,

$$y^2 = (x^2-a^2)(x^2-1)(x^2-b^2) \text{ and } y^2 = (x^2-c^2)(x^2-1)(x^2-d^2) \ .$$

In a sense we do not want to make precise, C can be considered as obtained by gluing a "half" of C' and a "half" of C''. The interesting fact about this informal observation is that experimental data shows that if,

$$\begin{pmatrix} y_1 & y_{12} \\ y_{12} & y_2 \end{pmatrix}$$

is a period matrix for C then the matrices,

$$\begin{pmatrix} y_1 & y_{12} \\ y_{12} & y_1 \end{pmatrix} \text{ and } \begin{pmatrix} y_2 & y_{12} \\ y_{12} & y_2 \end{pmatrix}$$

lie in the neighbourhood of period matrices of C' and C''. In practice this gives a way to get a first estimate on the coefficients a, b, c and d. And we can again conclude by adjusting coefficients using the same type of ideas as described before. In practice, since the curves only depend on 3 parameters we vary b, c and simultaneously a and d (this last choice has a mild consequence, namely that we must allow b or c to become negative).

5. Example

Various test versions of the algorithm have been implemented in C and are currently runing on both work stations and PC's. As an average the computations take a few minutes. We describe here how the method works on an example.

Let,

$$Z = \begin{pmatrix} i & i/4 \\ i/4 & 2i \end{pmatrix}$$

(note that, although this matrix has rational coefficients, it is non special in the sense that it does not correspond to a curve with non-trivial automorphisms).

We first compute approximations of curves with period matrices,

$$Z_1 = \begin{pmatrix} i & i/4 \\ i/4 & i \end{pmatrix} \text{ and } Z_2 = \begin{pmatrix} 2i & i/4 \\ i/4 & 2i \end{pmatrix}$$

using the first method. This yields the following sets of roots, $\pm.5435175550, \pm1$ and ±1.7029627189 for Z_1 and $\pm.9743781355, \pm1$ and ±2.1652082792 for Z_2. Hence our first approximation of the curve we are looking for is,

$$y^2 = (x + 1.7029627189)(x + 1)(x + .543517555)$$
$$(x - .9743781355)(x - 1)(x - 2.1652082792)$$

(here we are only illustrating the method, it is clear that in practice one does not need at this stage to compute the equations corresponding to Z_1 and Z_2 with such accuracy). A period matrix for this curve is approximately,

$$\begin{pmatrix} 1.0173788842\, i & 0.2478000968\, i \\ 0.2478000968\, i & 1.9822971468\, i \end{pmatrix}$$

which, as claimed is reasonably close to Z.

The next step is to adjust coefficients. It is unclear what is the best strategy one should use, the one we have used is based on the fact that perturbating b (resp. c) will act by a larger factor on y_1 (resp. y_2) and y_{12} than on y_2 (resp. y_1) while perturbating simultaneously a and d will act with a similar factor on all 3 coefficients. In this particular example our method does this adjustment in 32 steps, which is a relatively low value (it generally takes about a hundred).

The final equation returned by the program is,

$$y^2 = (x + 1.7006048665)(x + 1)(x + 0.514499336)$$
$$(x - 0.9758465301)(x - 1)(x - 2.1628514995)\ .$$

This curve has a period matrix whose coefficients coincide with those of Z up to 9 decimal points (which is the accuracy asked for).

The actual computation took 139 seconds on a 486 PC running at 50 Mhz with 8 Mega Bytes of RAM.

References

1. E. Bujalance, J.J. Etayo, J.M. Gamboa, G. Gromadski, *Automorphism groups of compact bordered Klein surfaces*, Springer Lecture Notes 1439, Berlin Heidelberg New York, 1990.
2. Y. Chekouri, *Thse*, In preparation.
3. D. Mumford, *Tata lectures on Theta* II, Birkäuser, Boston Basel Stuttgart, 1984.
4. A. Robert, *Elliptic Curves*, Springer Lecture Notes in Math 326, Berlin Heidelberg New York, 1973.
5. R. Silhol, *Compactifications of moduli spaces in real algebraic geometry*, Inventiones Math.,**107** (1992), 151–202.
6. R. Silhol, *Normal forms of period matrices of real algebraic curves of genus 2 and 3*, Journal of Pure and Applied Algebra, **87** (1993), 79–92,.
7. R. Silhol, *A numerical approach to computing period matrices of algebraic curves*, To appear.

Universit Montpellier II, Dpartement de Mathmatiques, URA CNRS 1407, Place E. Bataillon 34095 Montpellier Cedex 5, France

Contemporary Mathematics
Volume **182**, 1995

Counter-examples to Ragsdale Conjecture and T-curves

ILIA ITENBERG

Abstract. The paper is devoted to the construction of counter-examples to the Ragsdale conjecture, an old and an important conjecture in the topology of real algebraic curves. The constructed counter-examples are T-curves, i.e. curves which can be obtained by a combinatorial procedure based on Viro's method of construction of real algebraic varieties with prescribed topology. T-curves are also used in the paper to obtain a classification of M-curves of degree $4l+2$ with one non-empty oval, and to show the sharpness of the second Petrovsky inequality.

1. Introduction

In 1906 V. Ragsdale [8] analyzing the results of Harnack's and Hilbert's constructions proposed an important conjecture on the topology of real algebraic curves. To formulate it let us consider a real algebraic plane projective curve of even degree $m = 2k$, i.e. a real homogeneous polynomial of degree $2k$ in three variables defined up to multiplication by a non-zero real number. We suppose a curve to be non-singular, which means that a polynomial does not have singular points in $\mathbf{R}^3 \setminus 0$.

Such a curve A has a well defined zero locus $\mathbf{R}A$ in the real projective plane $\mathbf{R}P^2$. The set $\mathbf{R}A$ is a union of non-intersecting circles embedded in $\mathbf{R}P^2$. The topological type of the pair ($\mathbf{R}P^2$, $\mathbf{R}A$) is defined by the scheme of disposition of the components of $\mathbf{R}A$. This scheme is called *the real scheme of curve A*.

The real point set $\mathbf{R}A$ of the curve A divides the real projective plane $\mathbf{R}P^2$ in two parts with a common boundary $\mathbf{R}A$ (these parts are the subsets of $\mathbf{R}P^2$ where a polynomial has positive or, respectively, negative values). One of these parts is non-orientable, we will denote it by $\mathbf{R}P^2_-$. The other one will be denoted by $\mathbf{R}P^2_+$.

The topology of $\mathbf{R}P^2_-$ and $\mathbf{R}P^2_+$ is closely connected with the topological type of the pair ($\mathbf{R}P^2$, $\mathbf{R}A$). Let p be the number of connected components of

1991 *Mathematics Subject Classification.* Primary 14P25 ; Secondary 14H99.

$\mathbf{R}P^2_+$, and $n+1$ be the number of connected components of $\mathbf{R}P^2_-$ (exactly one component of $\mathbf{R}P^2_-$ is non-orientable).

The numbers p and n can be described in another way. Each connected component of the real point set $\mathbf{R}A$ of a curve of even degree is called *an oval.* It divides $\mathbf{R}P^2$ in two parts. We call the part homeomorphic to a disk *the interior* of the oval.

An oval of a curve is called *even* (resp. *odd*) if it lies inside of an even (resp. odd) number of other ovals of this curve.

It is easy to see that p is the number of even ovals of a curve, and n is the number of odd ovals.

The statement of the Ragsdale conjecture is the following : for any curve of degree $2k$

$$p \leq \frac{3k^2 - 3k + 2}{2}, \quad n + 1 \leq \frac{3k^2 - 3k + 2}{2}$$

Ragsdale also proposed the other conjecture :

$$p - n \leq \frac{3k^2 - 3k + 2}{2}, \quad n - p + 1 \leq \frac{3k^2 - 3k + 2}{2}$$

So, the first Ragsdale conjecture is a statement on the maximal possible number of connected components of $\mathbf{R}P^2_+$ and $\mathbf{R}P^2_-$. The second conjecture is a statement on the maximal value of Euler characteristic of $\mathbf{R}P^2_+$ and $\mathbf{R}P^2_-$.

In 1938 I. Petrovsky [7] proved the second Ragsdale conjecture (the inequalities of this conjecture are called now the Petrovsky inequalities) and also proposed a conjecture similar to the first one :

$$p \leq \frac{3k^2 - 3k + 2}{2}, \quad n \leq \frac{3k^2 - 3k + 2}{2}$$

In 1980 O. Viro [11] constructed curves of degree $2k$ with $n = \frac{3k^2-3k+2}{2}$ for any even $k \geq 4$. These curves are counter-examples to the original Ragsdale conjecture, but not to the conjecture of Petrovsky.

The following theorem gives counter-examples to the "corrected" Ragsdale conjecture (or to the conjecture of Petrovsky) (see also [3], [4]).

THEOREM 1.1. *For any integer number $k \geq 1$*

a) there exists a non-singular real algebraic plane projective curve of degree $2k$ with

$$p = \frac{3k^2 - 3k + 2}{2} + \left[\frac{(k-3)^2 + 4}{8}\right]$$

(where $[a]$ denotes the maximal integer not greater than a),

b) there exists a non-singular real algebraic plane projective curve of degree $2k$ with

$$n = \frac{3k^2 - 3k + 2}{2} + \left[\frac{(k-3)^2 + 4}{8}\right] - 1$$

The proof of this theorem will be given in the section 4.

The question on the sharp upper bound for the numbers p, n of curves of even degrees is still open. One can obtain an upper bound using Harnack and Petrovsky inequalities. However, the known examples are far from this estimation.

The counter-examples constructed in this paper are T-curves, i.e. the curves which can be obtained by a combinatorial procedure based on Viro's method of construction of real algebraic varieties (see, for example, [9], [10], [12], [13], [15]). The definition of T-curves is given in the section 2, some of their properties are discussed in the section 7.

We give two other applications of T-curves : classification of M-curves (i.e. curves having the maximal possible number of connected components of the real point set for a given degree) of degree $4l+2$ with one non-empty oval (in the section 5, see also [4]), and construction of examples of curves showing the sharpness of the second Petrovsky inequality (in the section 6).

I would like to thank S. Akbulut, V. Kharlamov and O. Viro for the useful discussions.

2. T-curves

Let m be a positive integer number and T be the triangle

$$\{(x,\ y) \in \mathbb{R}^2,\ \ x \geq 0,\ \ y \geq 0,\ \ x+y \leq m\}.$$

Suppose that T is triangulated in such a way that the vertices of the triangles are integer, and that some distribution of signs, $a_{i,j} = \pm$ at the vertices of the triangulation, is given. Then there arises a naturally associated piecewise-linear curve L in $\mathbf{R}P^2$.

The construction of L is the following.

Take copies $T_x = s_x(T)$, $T_y = s_y(T)$, $T_{xy} = s(T)$ of T, where $s = s_x \circ s_y$ and s_x, s_y are reflections with respect to the coordinate axes. Extend the triangulation of T to a symmetric triangulation of $T \cup T_x \cup T_y \cup T_{xy}$ and extend the distribution of signs to a distribution at the vertices of the extended triangulation which verifies the modular property: $g^*(a_{i,j}x^iy^j) = a_{g(i,j)}x^iy^j$ for $g = s_x,\ s_y,\ s$ (other words, the sign at a vertex is the sign of the corresponding monomial in the quadrant containing the vertex).

If a triangle of the triangulation has vertices of different signs, select a midline separating them. Denote by L' the union of the selected midlines. It is contained in $T \cup T_x \cup T_y \cup T_{xy}$. Glue by s the sides of $T \cup T_x \cup T_y \cup T_{xy}$. The resulting space T_* is homeomorphic to $\mathbf{R}P^2$. Let us take the curve L to be the image of L' in T_*.

A pair (T_*, L) is called *a chart* of a real algebraic plane projective curve A, if there exists a homeomorphism of pairs $(T_*, L) \longrightarrow (\mathbf{R}P^2, \mathbf{R}A)$.

Let us introduce two additional assumptions : the considered triangulation of T is *primitive* and *convex*. The first condition means that all triangles are of area $1/2$ (or, equivalently, that all integer points of T are vertices of the triangulation). The second one means that there exists a convex piecewise-linear function $T \longrightarrow \mathbf{R}$ which is linear on each triangle of the triangulation and not linear on the union of two triangles.

The following statement is the special case of Viro's theorem [13, Th. 1.4].

THEOREM 2.1 (O. VIRO). *Under the assumptions made above on the triangulation of the triangle T, there exists a non-singular real algebraic plane projective curve A of degree m with the chart (T_*, L).*

A curve having the chart (T_*, L) is called *a T-curve.* This notion was introduced by S. Orevkov [6].

3. Construction of Harnack curves

Recall that an M-curve is a curve having the maximal possible number of connected components of the real point set for a given degree. Harnack [2] proved that this maximal number is equal to $\frac{(m-1)(m-2)}{2} + 1$ for the degree m.

In this section we will describe, using Theorem 2.1, the construction of some M-curves (a special case of Harnack curves). This construction will play an important role in the sections 4, 5, 6.

Let $m = 2k$ be a positive even number, and T again be the triangle

$$\{(x,\ y) \in \mathbb{R}^2,\ \ x \geq 0,\ \ y \geq 0,\ \ x + y \leq m\}.$$

An integer point (i, j) of T is called *even,* if i, j are both even, and *odd* if not.

The following distribution of signs at the integer points of the triangle T is called *Harnack distribution* :

even points get sign "-", and odd points get sign "+".

We use the system of notations for the real schemes of non-singular curves suggested by Viro [11]. The scheme consisting of a single oval is denoted by the symbol $< 1 >$, the empty scheme - by the symbol $< 0 >$. If a symbol $< A >$ stands for some set of ovals, then the set of ovals obtained by addition of an oval surrounding all old ovals is denoted by $< 1 < A >>$. If a scheme is the union of two non-intersecting sets of ovals denoted by $< A >$ and $< B >$ respectively with no oval of one set surrounding an oval of the other set, then this scheme is denoted by the symbol $< A \cup B >$. Besides, if A is the notation for some set of ovals then a part $A \cup \ \dots\ \cup A$ of another notation where A repeats n times is denoted by $n \times A$; a part $n \times 1$ is denoted by n.

PROPOSITION 3.1. *An arbitrary primitive convex triangulation of T with the Harnack distribution of signs at the vertices produces a T-curve which is an M-curve of degree $m = 2k$ with the real scheme*

$$< \frac{3k^2 - 3k}{2} \cup 1 < \frac{(k-1)(k-2)}{2} >>$$

PROOF. Let us, first, notice that the number of interior (i.e. lying strongly inside of the triangle T) integer points is equal to $\frac{(m-1)(m-2)}{2}$, the number of even interior points is equal to $\frac{(k-1)(k-2)}{2}$, and the number of odd interior points is equal to $\frac{3k^2-3k}{2}$.

If an interior vertex (i, j) is even, then its star $St(i, j)$ contains an oval of the curve L (this oval surrounds "-"). If an interior vertex (i, j) is odd, then one of the symmetric copies of the star $St(i, j)$ (namely, the copy in the quadrant $\{(x, y) \mid sign(x) = (-1)^j,\ sign(y) = (-1)^i\}$) contains an oval of L (this oval surrounds "+").

We have found $\frac{(m-1)(m-2)}{2}$ ovals and, thus, the curve L can have only one oval more. This oval exists, because, for example, the curve L intersects the coordinate axes.

To finish the proof it remains to notice that the union of the segments

$$\{x - y = -m,\ -m \leq x, y \leq m\} \cup$$

$$\{x \leq 0,\ y = 0,\ -m \leq x, y \leq m\} \cup \{x = 0,\ y \leq 0,\ -m \leq x, y \leq m\}$$

is not contractable in T_* and contains only the signs "-". It means that $\frac{3k^2-3k}{2}$ ovals corresponding to odd interior points and containing the sign "+" inside of them are situated outside of the non-empty oval. □

4. Construction of counter-examples to Ragsdale conjecture

Now we are able to describe a construction of counter-examples to Ragsdale conjecture.

PROOF OF THEOREM 1.1. We will construct T-curves with the stated properties. Let us show, first, how to construct a curve of degree $m = 2k$ with $p = \frac{3k^2-3k+2}{2} + 1$.

Suppose that the hexagon S shown in Figure 1 is placed inside of the triangle $T = \{x \geq 0,\ \ y \geq 0,\ \ x + y \leq m\}$ in such a way that the center of S has both coordinates odd. Any convex primitive triangulation of a convex part of a convex polygon is extendable to a convex primitive triangulation of the polygon. Let us extend to T the convex primitive triangulation of S. Extend also the distribution of signs using the Harnack distribution outside of S.

The corresponding piecewise-linear curve L has $\frac{3k^2-3k+2}{2} + 1$ even ovals, and its real scheme is the following :

$$< \frac{3k^2 - 3k - 2}{2} \cup 1 < \frac{(k-1)(k-2) - 8}{2} \cup 1 < 2 >>>$$

Suppose now that each marked hexagon presented in Figure 2 has the triangulation and the signs of S. The triangulation of the union $\widetilde{S}$ of the marked hexagons can be extended to a primitive convex triangulation of T. Let us fix such an extension. Outside of $\widetilde{S}$ choose again the Harnack distribution of signs at the vertices of the triangulation.

One can calculate that for the corresponding piecewise-linear curve L we have

$$p = \frac{3k^2 - 3k + 2}{2} + a$$

where a is the number of the marked hexagons, and $a = \left[\frac{(k-3)^2+4}{8}\right]$.

The constructed curve has the following real scheme

$$< \frac{3k^2-3k-2a}{2} \cup 1 < \frac{(k-1)(k-2)-8a}{2} \cup a \times 1 < 2 >>>$$

To prove the part b) of the theorem, let us fix in addition to the triangulation of $\widetilde{S}$ a triangulation of some part P of a neighbourhood of the axe OY and the signs at the vertices of this triangulation as it is shown in Figure 3 (more precisely, only the case $k \equiv 1 \ mod\ 4$ is presented in this figure, if $k \not\equiv 1 \ mod\ 4$ one should change the triangulation near the point $(0, m)$). The chosen primitive convex triangulation of $\widetilde{S} \cup P$ can be extended to a primitive convex triangulation of the triangle T. Outside of $\widetilde{S} \cup P$, let us take the Harnack distribution of signs at the vertices of the triangulation.

For the corresponding piecewise-linear curve L

$$n = \frac{3k^2-3k+2}{2} + \left[\frac{(k-3)^2+4}{8}\right]$$

(the case $k \equiv 1 \ mod\ 4$) or

$$n = \frac{3k^2-3k+2}{2} + \left[\frac{(k-3)^2+4}{8}\right] - 1$$

(the case $k \not\equiv 1 \ mod\ 4$). □

Recently, B. Haas [1] modified the presented construction and obtained examples of T-curves of degree $2k$ with

$$p = \frac{3k^2-3k+2}{2} + \left[\frac{k^2-7k-10}{6}\right]$$

5. M-curves with one non-empty oval

In this subsection we discuss a classification of the real schemes of M-curves of degree $4l+2$ with one non-empty oval.

Let us start with two well-known restrictions for the topology of real plane projective curves (see, for example, [14], [16]).

Gudkov - Rokhlin congruence

$p - n \equiv k^2 \ mod\ 8$ *for an M-curve of degree* $2k$

Improved Petrovsky inequalities

Let A be a curve of degree $2k$. Denote by p_- (resp. by n_-) the number of even (resp. odd) ovals of $\mathbf{R}A$ *bounding from the exterior the components of* $\mathbf{R}P^2 \setminus \mathbf{R}A$ *with negative Euler characteristic. Then*

$$p - n_- \leq \frac{3k^2-3k+2}{2}, \quad n - p_- + 1 \leq \frac{3k^2-3k+2}{2}$$

Improved Petrovsky inequalities for M-curves of degree $4l+2$ with one non-empty oval can be rewritten (using the Gudkov - Rokhlin congruence and the fact that $p_- \leq 1,\ n_- = 0$ in this case) as follows :

$$p \leq \frac{3k^2 - 3k + 2}{2}, \quad n \leq \frac{3k^2 - 3k}{2}$$

It turns out that the Gudkov - Rokhlin congruence and the improved Petrovsky inequalities are the only restrictions for the topology of M-curves of degree $4l+2$ with one non-empty oval.

THEOREM 5.1. *Let $m = 2k = 4l + 2$, where l is a positive integer number. Then for any positive integer numbers p, n such that*

$$p + n = \frac{(m-1)(m-2)}{2} + 1$$

satisfying the Gudkov - Rokhlin congruence and the improved Petrovsky inequalities there exists a real algebraic plane projective M-curve of degree m with the real scheme

$$<(p-1) \cup 1 < n >>$$

PROOF. Let $\varepsilon, \delta \in \{0,1\}$. Denote by $H^+_{\varepsilon,\delta}$ (resp. $H^-_{\varepsilon,\delta}$) the following distribution of signs at integer points :

a vertex (i,j) gets sign "+" (resp. "-"), if $i \equiv \varepsilon \bmod 2,\quad j \equiv \delta \bmod 2$, and sign "-" (resp. "+") otherwise.

So, the Harnack distribution of signs, described in the section 3, is denoted now by $H^-_{0,0}$.

Remark that any distribution $H^\pm_{\varepsilon,\delta}$ can be formulated as the Harnack one for the appropriate quadrant of the plane (exchanging, if necessary, "+" and "-"). Thus, Proposition 3.1 also holds true for any distribution $H^\pm_{\varepsilon,\delta}$.

Let us divide the triangle T in two polygons T_1 and T_2 (where T_1 is a quadrangle, T_2 is a triangle) by a segment with the following properties :

(i) the ends of the segment are odd points lying on the boundary of T,
(ii) the segment does not contain integer points except the ends.

Consider an arbitrary convex primitive triangulation in each polygon T_1, T_2 (the union of these triangulations is a convex primitive triangulation of T, because the chosen segment does not contain vertices of the triangulations except the ends). Let us take the Harnack distribution of signs in T_1 and choose in T_2 the only distribution $H^-_{\varepsilon,\delta}$ different from the Harnack one and compatible on the common boundary of the polygons with the chosen distribution in T_1.

The arguments of the proof of Proposition 3.1 show again that the triangulation and the distribution of signs described above give an M-curve with one non-empty oval.

Let P_1, P_2 be the numbers of interior even points of T_1 and T_2, and N_1, N_2 be the numbers of interior odd points of these polygons.

One can see that for the curve obtained

$$p = N_1 + P_2 + 1, \quad n = P_1 + N_2$$

To prove Theorem 5.1, we divide the triangle T by segments with the properties (i), (ii) as it shown in Figure 4 (the vertices on the axe OX have the coordinates $(k + (4i+2),\ 0)$, the vertices on the axe OY have the coordinates $(0,\ k + (4i+2))$, and the vertices on the line $x + y = m$ have the coordinates $(k \pm 4i,\ k \mp 4i)$ with appropriate values of a non-negative integer i).

Let us take in the quadrangle $OABC$ the Harnack distribution of signs. Now we choose one of the distributions $H^-_{0,0}$, $H^-_{0,1}$ in each triangle part of the subdivision of T lying under the line $x = y$. Also, we choose one of the distributions $H^-_{0,0}$, $H^-_{1,0}$ in each triangle part of the subdivision of T lying over the line $x = y$.

Finally, we choose in each part an arbitrary convex primitive triangulation (actually, the real scheme of the resulting T-curve does not depend of this choice of primitive triangulations). The chosen triangulation and distribution of signs produce an M-curve with one non-empty oval. It is easy to verify that all possible (in the sense of the statement of Theorem 5.1) pairs p, n can be realized using the described construction. For example, to realize two extremal cases one can take the Harnack distribution of signs in T (the case $p = \frac{3k^2-3k+2}{2}$, $n = \frac{(k-1)(k-2)}{2}$, a Harnack curve) or the Harnack distribution in the quadrangle $OABC$ and the distributions $H^-_{0,1}$ and $H^-_{1,0}$ in $T \setminus OABC$ (for the opposite extremal case $p = \frac{(k-1)(k-2)}{2} + 1$, $n = \frac{3k^2-3k}{2}$). □

6. Sharpness of the second Petrovsky inequality

It is well-known that the first Petrovsky inequality

$$p - n \leq \frac{3k^2 - 3k + 2}{2}$$

is sharp for any positive integer k. One can take, for example, a Harnack curve of degree $2k$ with the real scheme

$$< \frac{3k^2 - 3k}{2} \cup 1 < \frac{(k-1)(k-2)}{2} >>$$

and contract all odd ovals. Let us explain how this contraction can be done for T-curves described in the section 3. To contract an odd oval α in this case, one can choose such a triangulation of the triangle T that the corresponding to α interior even vertex would have exactly three neighbours, and then change the sign of this vertex.

The sharpness of the second Petrovsky inequality

$$n - p + 1 \leq \frac{3k^2 - 3k + 2}{2}$$

is more difficult to show. A curve with $n - p + 1 = \frac{3k^2-3k+2}{2}$ is, evidently, a counter-example to Ragsdale conjecture.

PROPOSITION 6.1. *The second Petrovsky inequality*

$$n - p + 1 \leq \frac{3k^2 - 3k + 2}{2}$$

is sharp for any integer $k \geq 4$.

PROOF. To obtain the curves with

$$n - p + 1 = \frac{3k^2 - 3k + 2}{2}$$

for an even $k \geq 4$, one can take a curve constructed by Viro (see [11]) with the real scheme

$$< \frac{k^2 - 3k}{2} \cup 1 < \frac{3k^2 - 3k + 2}{2} >>$$

and contract all even empty ovals (it is possible in Viro's construction).

Suppose now, that k is odd and $k \geq 5$, $k \neq 7$ (the case $k = 7$ we will consider later). We can choose a positive number a and a non-negative number b such that $6a + 8b + 4 = 2k$. Let us consider a partition of the triangle T shown in Figure 5, where the centers of the hexagons $S_1, ..., S_a, S_{a+1}, ..., S_{a+b}$ have the second coordinates

$$3, ..., 6(a-1) + 3,\ 6(a-1) + 11, ..., 6(a-1) + 8b + 3$$

respectively. Suppose that the triangulation and the signs of each marked hexagon coincide with the triangulation and the signs of the hexagon S (shown in Figure 1). The chosen primitive convex triangulation of the union $\widetilde{S}$ of the marked hexagons and of the part P' of a neighbourhood of the axe OY can be extended to a primitive convex triangulation of the triangle T. Moreover, this extension can be chosen in such a way that any interior even vertex of T lying in $T \setminus (\widetilde{S} \cup P')$ would have exactly three neighbours. Now, let us put the sign "+" at all integer points of $T \setminus (\widetilde{S} \cup P')$ except the even points lying on the boundary of T, where we put the sign "-".

The corresponding T-curve has the real scheme

$$< (a+b) \times 1 < 2 > \cup 1 < \frac{3k^2 - 3k + 2}{2} - (a+b) >>$$

(in the case $k = 9$ the oval of the T-curve intersecting the coordinate axes is shown in Figure 6).

Consider now the case $k = 7$. Take the partition of T shown in Figure 7. We choose inside of the hexagon S_1 the triangulation and the signs of S. The triangulation of S_1 and of the polygon P'' with the vertices $(0,5)$, $(7,7)$, $(8,6)$, $(6,5)$, $(5,6)$ can be extended to a convex primitive triangulation of T in such a way that any interior even vertex lying outside of S_1 and P'' would have three neighbours.

Let us take the Harnack distribution of signs at the integer points of the quadrangle $\{(0,0), (0,5), (7,7), (14,0)\}$ except the points of S_1 and P''. In the triangle $\{(0,5), (0,14), (7,7)\}$ we choose the distribution $H^+_{0,1}$ (see section 5).

The corresponding T-curve has the real scheme $< 1 < 2 > \cup 1 < 63 >>$. It rests to contract even empty ovals. It can be done changing "-" for "+" at the interior even points of $T \setminus (S_1 \cup P'')$. □

7. How large is the class of T-curves ?

It is natural to pose the following question : can the real scheme of an arbitrary non-singular real algebraic plane projective curve be realized by a T-curve of the same degree ?

One can immediately find a trivial restriction : evidently, the empty real scheme of a curve of an even degree cannot be realized by T-curves. We will formulate another restriction.

Let us, first, give a necessary definition. A pair of ovals is called *injective* if one oval of this pair lies inside of the other one. Denote by J the number of ovals of a curve containing inside of them at least one injective pair.

THEOREM 7.1. *For a T-curve of degree m being an M-curve the following inequality holds*

$$J \leq 3m/2$$

To prepare the proof of Theorem 7.1 we will prove the Harnack inequality (the number of components of the real point set of a curve of degree m is not greater than $\frac{(m-1)(m-2)}{2} + 1$) for T-curves in a combinatorial way.

Take an arbitrary primitive triangulation of the triangle T with some distribution of signs at the vertices. We will show that the number of connected components of $T_* \setminus L$ is at most $\frac{(m-1)(m-2)}{2} + 2$.

Consider a sequence $U_1, U_2, \dots, U_i, \dots$, where each U_i is a union of some triangles of the triangulation of T, the element U_1 consists of one triangle with the side [(0;0),(1;0)], and U_i is defined by induction as follows. Denote by γ_i an edge of the triangulation lying on the boundary of U_{i-1} but not on the boundary of the triangle T (if such an edge does not exist, then U_{i-1} coincides with T and the construction of the sequence is finished). Let Γ_i be the triangle of the triangulation which does not belong to U_{i-1} and has the side γ_i. Let v_i be the vertex of Γ_i which does not belong to γ_i. Then we take U_i equal to the union of all triangles of U_{i-1} and of all triangles Γ such that v_i is a vertex of Γ, and two other vertices of Γ belong to U_{i-1}.

Now, consider the symmetric copies

$$U_{i,x} = s_x(U_i), \quad U_{i,y} = s_y(U_i), \quad U_{i,xy} = s(U_i)$$

of U_i, where $s = s_x \circ s_y$ and s_x, s_y again are reflections with respect to the coordinate axes. Let $V_i = U_i \cup U_{i,x} \cup U_{i,y} \cup U_{i,xy}$. Denote by $C(i)$ the number of connected components of $T_* \setminus L$ intersecting V_i.

The following statements are easy to verify :

i) $C(1) \leq 3$, and $C(1) \leq 2$ if all vertices of U_1 belong to the boundary of T,
ii) $C(i) \leq C(i-1) + 1$,
iii) $C(i) \leq C(i-1)$, if the vertex v_i belongs to the boundary of T.

The number of interior integer points of the triangle T is equal to $\frac{(m-1)(m-2)}{2}$. So, the number of connected components of $T_* \setminus L$ is at most $\frac{(m-1)(m-2)}{2} + 2$.

One can obtain the statement below using the described combinatorial proof of the Harnack inequality for T-curves.

PROPOSITION 7.2. *Let a convex primitive triangulation of the triangle T and a distribution of signs at the vertices of this triangulation be chosen in such a way that the resulting T-curve is an M-curve. If the star $St(v)$ of an interior vertex v does not intersect the boundary of T, then $St(v)$ or one of its symmetric copies contains an oval of the curve L.*

PROOF. Let us construct a sequence $U_1, \dots, U_r$ (where $r = \frac{(m+1)(m+2)}{2} - 2$) as it was described in the proof of the Harnack inequality but verifying the additional condition, that the last vertex v_r coincides with v (it is possible, because $St(v)$ does not intersect the boundary of T).

Our T-curve is an M-curve, thus $C(i)$ should be equal to $C(i-1)+1$ if v_i is an interior vertex. In particular, we have $C(r) = C(r-1)+1$, which means that the star $St(v)$ or one of its symmetric copies contains an oval of L. □

The statement of Proposition 7.2 was, first, proved by V. Kharlamov [5] in a different (non-combinatorial) way.

PROOF OF THEOREM 7.1. We will show that for a T-curve being an M-curve any oval of the curve L surrounding an injective pair should intersect the boundary of one of the triangles T, $s_x(T)$, $s_y(T)$, $s(T)$. It will give the proof of the theorem, because the number of the points of intersection of L and the boundary of these triangles is equal to $3m$.

Suppose that β is an oval of L containing inside an injective pair and non-intersecting the boundary of T, $s_x(T)$, $s_y(T)$, $s(T)$. Let β lie, for example, inside of T. Denote by β_1 the exterior oval of the injective pair. All vertices inside of β_1 have stars non-intersecting the boundary of T. Let W be the union of the stars of all vertices lying inside of β_1 and let w be the number of interior points of W. Then W and its symmetric copies contain at least $w+1$ ovals of L (one for each interior point of W as it was proved in Proposition 7.2, and also the oval β_1).

Consider a sequence $U_1, U_2, \dots, U_r$ described in the proof of the Harnack inequality. Let i be the minimal number such that v_i lies inside of W. It is easy to see that $C(i)$ should be greater than $C(i-1)+1$. This contradiction proves the theorem. □

Remark that the Theorem 7.1 gives a strong restriction on the topology of T-curves being M-curves. One can easily construct such a family of M-curves of increasing degrees that the numbers J of the curves of this family would depend quadratically in the degree.

The following question is open : is it true that a T-curve of degree m being an M-curve has no more than $O(m)$ non-empty ovals ?

REFERENCES

1. B. Haas, *Les multilucarnes : Nouveaux contre-exemples à la conjecture de Ragsdale*, (to appear in C. R. Acad. Sci.).
2. A. Harnack, *Über Vieltheiligkeit der ebenen algebraischen Curven*, Math. Ann. **10** (1876), 189 - 199.
3. I. Itenberg, *Contre-exemples à la conjecture de Ragsdale*, C. R. Acad. Sci. Paris **317, Série I** (1993), 277 - 282.
4. I. Itenberg, *Viro's method and T-curves*, to appear in MEGA, Santander, 1994.
5. V.Kharlamov, *Private communication.*

6. S. Orevkov, *Private communication.*
7. I. Petrovsky, *On the topology of real plane algebraic curves*, An. Math. **39** (1938), no. 1, 187 - 209.
8. V. Ragsdale, *On the arrangement of the real branches of plane algebraic curves*, Amer. J. Math. **28** (1906), 377 - 404.
9. J.-J. Risler, *Construction d'hypersurfaces réelles (d'après Viro)*, Séminaire N. Bourbaki (1992), no. 763.
10. O. Viro, *Construction of multicomponent real algebraic surfaces*, Soviet Math. Dokl. **20** (1979), no. 5, 991 - 995.
11. O. Viro, *Curves of degree 7, curves of degree 8, and Ragsdale conjecture*, Soviet Math. Dokl. **22** (1980), 566 - 569.
12. O. Viro, *Gluing of algebraic hypersurfaces, smoothing of singularities and construction of curves*, Proc. Intern. Topol. Conf., Nauka, Leningrad, 1983, pp. 149 -197. (Russian)
13. O. Viro, *Gluing of plane real algebraic curves and construction of curves of degree 6 and 7*, Lecture Notes in Mathematics, vol. 1060, Springer, 1984, pp. 187 - 200.
14. O. Viro, *Progress in the topology of real algebraic varieties over the last six years*, Rus. Math. Surv. **41** (1986), no. 3, 55 - 82.
15. O. Viro, *Real algebraic plane curves : constructions with controlled topology*, Leningrad Math. J. **1** (1990), 1059 - 1134.
16. G.Wilson, *Hilbert's sixteenth problem*, Topology **17** (1978), no. 1, 53 - 73.

IRMAR, Université de Rennes 1, Campus de Beaulieu, F-35042 Rennes cedex, France

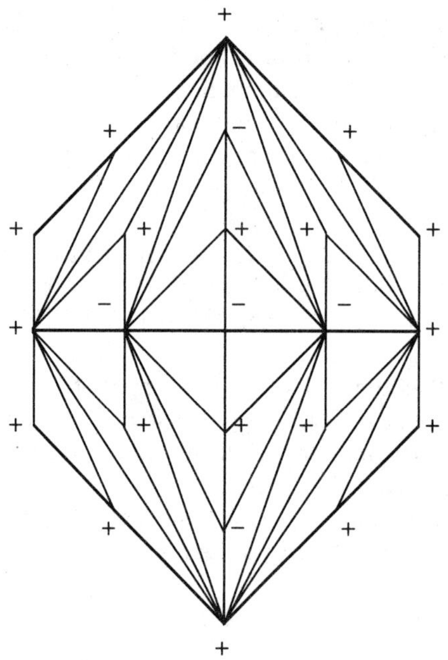

Figure 1. Hexagon S

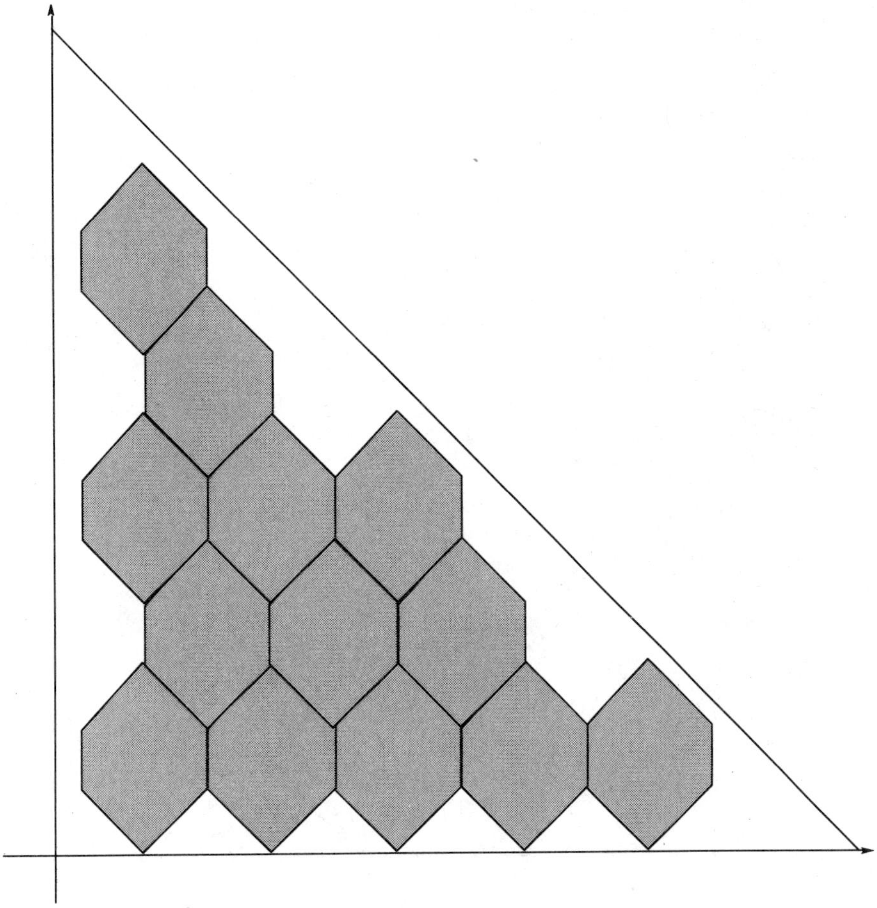

Figure 2. Partition for part a) of Theorem 1.1

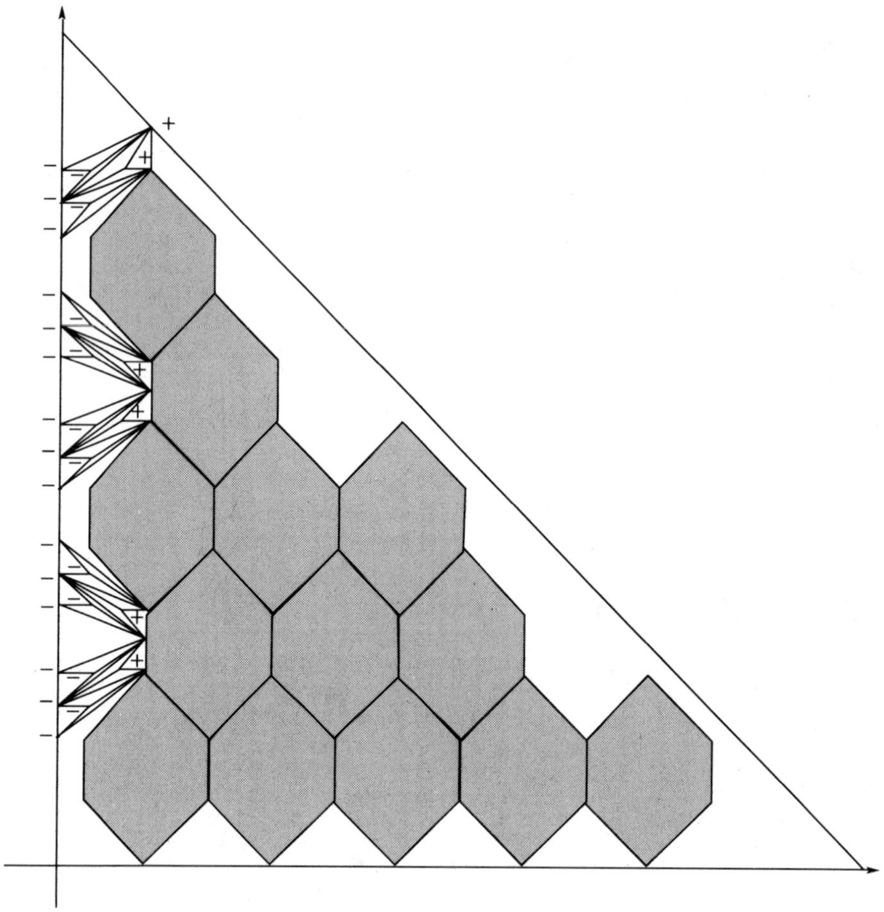

Figure 3. Partition for part b) of Theorem 1.1

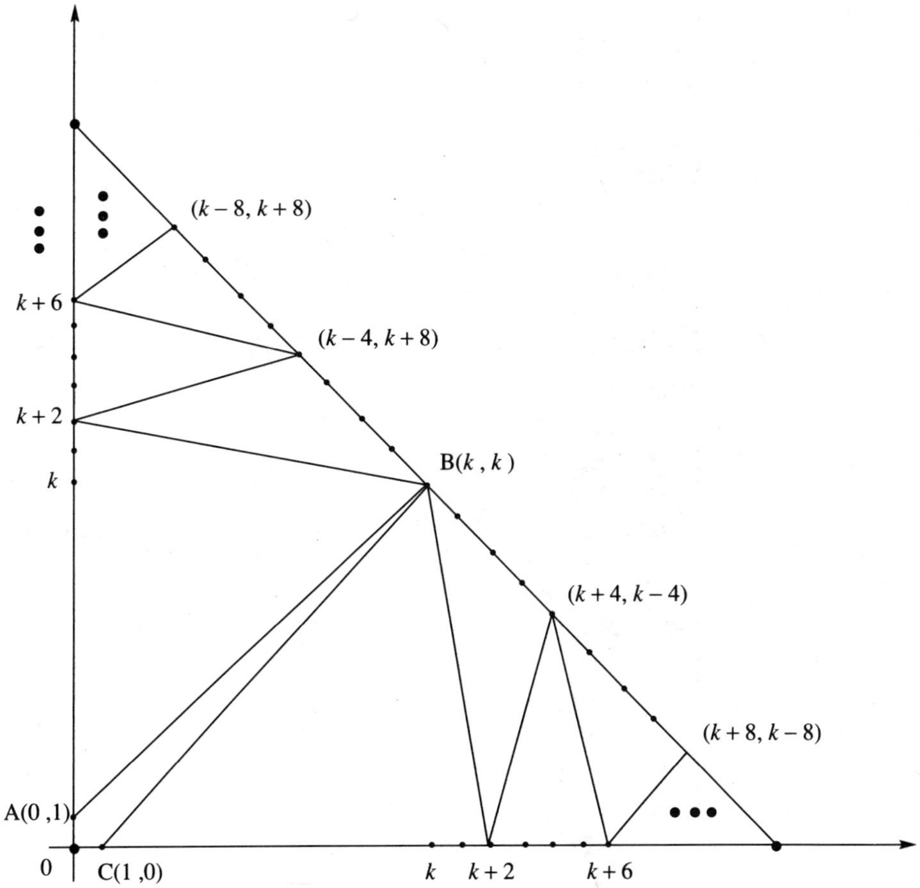

Figure 4. Partition for Theorem 5.1

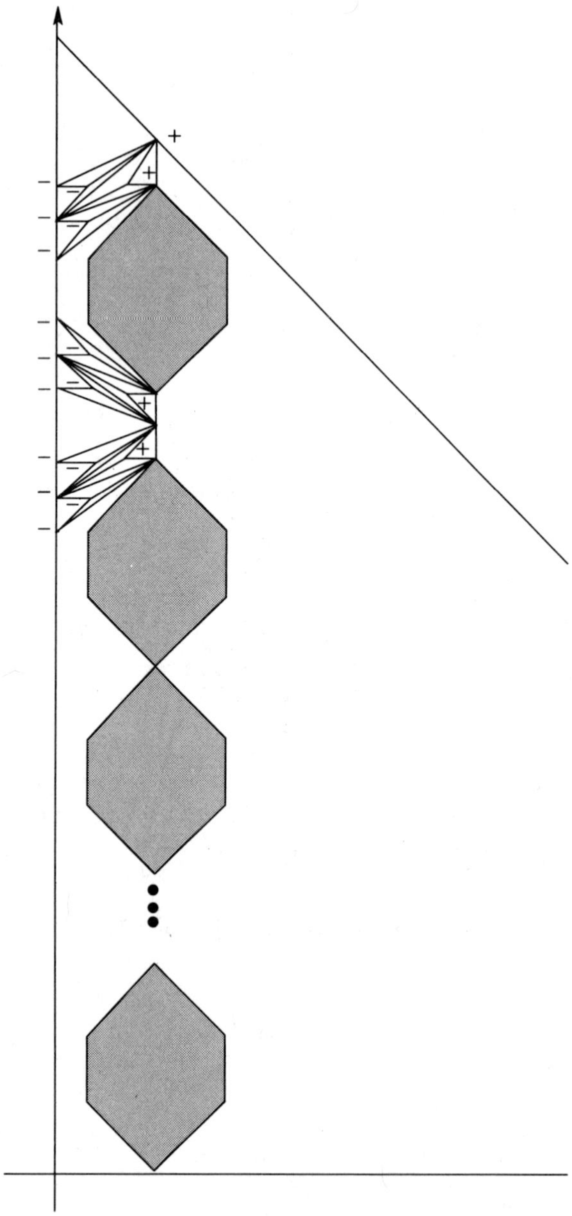

Figure 5. Partition for Proposition 6.1

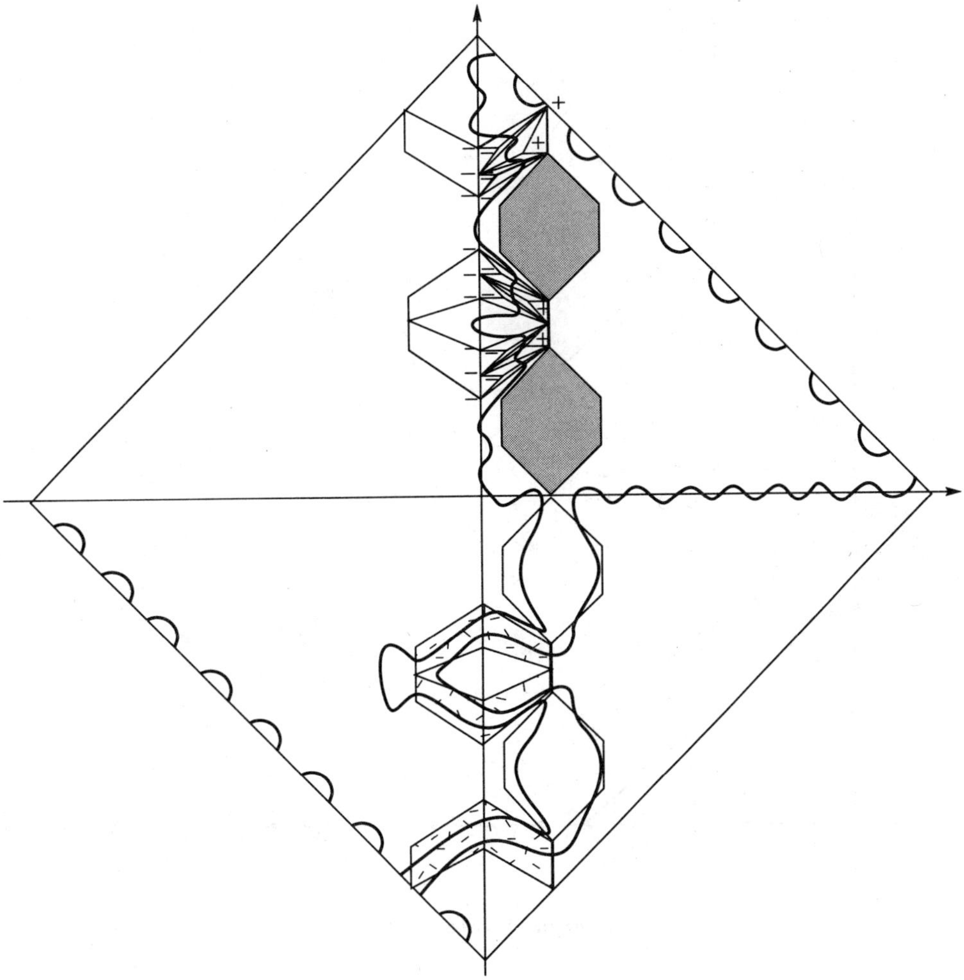

Figure 6. Case $k = 9$

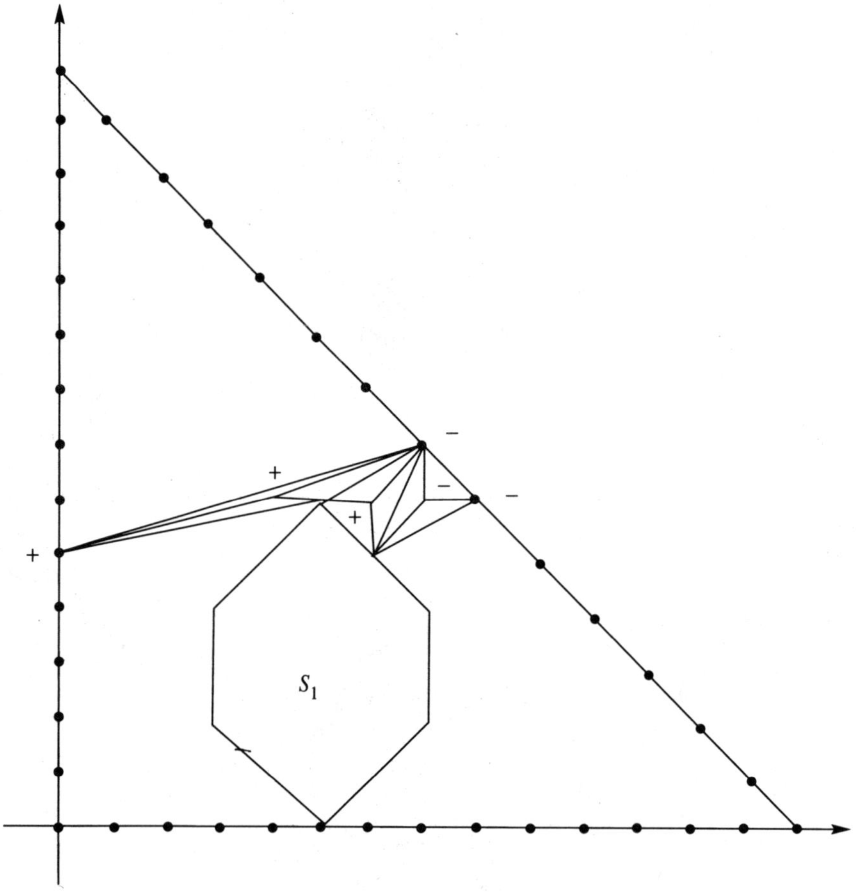

Figure 7. Case $k = 7$

Contemporary Mathematics
Volume **182**, 1995

Surfaces of small genus in connected sums of $\mathbb{C}P^2$ and real algebraic curves with many nests in $\mathbb{R}P^2$

G. MIKHALKIN

ABSTRACT. The subject of this paper is construction of smooth embedded surfaces of small genus realizing a fixed integer 2-dimensional homology class of a smooth 4-manifold. I show also that the question of finding the exact lower bound for the genus of a surface representing a fixed 2-dimensional homology class of $\mathbb{C}P^2$ (the Thom conjecture) makes sense even after stabilization. The stabilization of this question may appear to be simpler than the initial question (general situation for smooth 4-topology). I do not know the complete answer for the stabilization of the minimal genus problem for $\mathbb{C}P^2$ completely, but, in this paper, I show that stabilization allows to improve the known upper bound for the minimal genus of surfaces in $\mathbb{C}P^2$ (coming from holomorphic curves) while the lower bound coming from Rokhlin-Hsiang-Szczarba inequality still holds after stabilization. Some real algebraic curves in $\mathbb{R}P^2$ appear to be useful for construction of surfaces of small genus in connected sums.

1. Introduction

Let M be a smooth closed 4-manifold. Any homology class $x \in H_2(M)$ is representable by a smooth embedded orientable surface, say $F \subset M$. Let $g(F)$ be the genus of F. Then, x is representable by a smooth surface of genus $g(F)+1$, to see this it suffices to add a small trivial handle to F. However, it is not clear in general if x is representable by a surface of genus $g(F)-1$. The minimal value $g(x)$ of the genus of a surface representing x is an invariant of homology class x and this invariant is the main subject of this paper.

Even if M is simply-connected $g(x)$ is not always zero as it was observed by Rokhlin, Kervaire and Milnor. The smallest degree of a homology class in $\mathbb{C}P^2$ not representable by an embedded sphere is 3,

$$g(3[\mathbb{C}P^1] \in H_2(\mathbb{C}P^2)) \geq 1,$$

1991 *Mathematics Subject Classification.* Primary 57R95, 57M99; Secondary 14P25.

Supported by the NSF grant DMS 9304580 to the Institute for Advanced Study.

this follows from the Rokhlin theorem on signature, since $3[\mathbb{C}P^1]$ is a characteristic homology class and its self-intersection is not congruent modulo 16 to the signature of $\mathbb{C}P^2$.

In this paper we deal with the simply connected case.

REMARK 1.1. If M is not simply connected, there can be a homology lower bound for $g(x)$. Namely, consider the homomorphism of cap product with x

$$x\cap : H^1(M;\mathbb{R}) \to H_1(M;\mathbb{R}).$$

The rank of this homomorphism provides a lower bound for $g(x)$:

$$g(x) \geq \frac{\mathrm{rk}(x\cap)}{2},$$

since for any surface $F \subset M$ realizing x every element of the image of $x\cap$ can be realized geometrically as the intersection of F and some codimension 1 submanifold of M and, therefore,

$$\dim H_1(F;\mathbb{R}) \geq \mathrm{rk}(x\cap).$$

Acknowledgment. I would like to thank S.Akbulut, R.Gompf and O.Viro for useful discussions. I would like to thank the Institute for Advanced Study for hospitality during the research.

2. Real plane algebraic curves with many nests

Let $\mathbb{R}A$ be a nonsingular real algebraic curve of degree m in $\mathbb{R}P^2$. Topologically, $\mathbb{R}A$ is a disjoint union of circles. A circle in $\mathbb{R}P^2$ is either isotopic to $\mathbb{R}P^1$ or bounds a disk in $\mathbb{R}P^2$. It is easy to see that if m is even then every component of $\mathbb{R}A$ bounds a disk in $\mathbb{R}P^2$ and if m is odd then every component but one bounds a disk in $\mathbb{R}P^2$. If a circle bounds a disk in $\mathbb{R}P^2$ we call it an *oval* and we call the disk the *interior* of the circle.

DEFINITION. We call a subset $\mathfrak{N}$ of ovals of $\mathbb{R}A$ a *nest* if for every two ovals of $\mathfrak{N}$ one of them sits in the interior of another.

The number of elements of $\mathfrak{N}$ is called the depth of the nest $\mathfrak{N}$. Two nests are called disjoint if the interior of any oval of one of the nests does not intersect with the second nest. For a union of disjoint nests we call the sum of the depths of the nests in the union the *total depth* of the union of nests.

The following lemma establishes a connection between real algebraic curves in $\mathbb{R}P^2$ and surfaces in $\#\mathbb{C}P^2$ which we exploit in this paper.

LEMMA 1.

a). *If there exists a nonsingular curve $\mathbb{R}A$ in $\mathbb{R}P^2$ of degree m with n disjoint nests of total depth d then there exists an embedded orientable surface of genus $\frac{(m-1)(m-2)}{2} - d + 1$ in $\#(n+1)\mathbb{C}P^2$ realizing $(m, 0, \dots, 0) \in H_2(\#(n+1)\mathbb{C}P^2)$.*

b). *If there exists a component of $\mathbb{R}A$ not contained in the union of n nests then there exists an embedded orientable surface of genus $\frac{(m-1)(m-2)}{2} - d$ in $\#(n+1)\mathbb{C}P^2$ realizing $(m, 0, \dots, 0) \in H_2(\#(n+1)\mathbb{C}P^2)$.*

Now we need to find suitable real curves to make Lemma 1 work (as we shall see in the next section for computing g_{st} we want to maximize $d-n$).

LEMMA 2. *If there exists a real plane nonsingular algebraic curve of degree m with n nests of total depth d then for any positive integer number s there exists a real plane nonsingular algebraic curve of degree sm with n nests of total depth sd.*

We can even increase the total depth of the disjoint nests in the case when there is a line in $\mathbb{R}P^2$ suitably situated with respect to the curve.

LEMMA 3. *If there exists a real plane nonsingular algebraic curve $\mathbb{R}A$ of degree m with n nests of total depth d and a line L in $\mathbb{R}P^2$ intersecting $\mathbb{R}A$ in m points and intersecting each oval of the n nests in not more than 2 points then for any positive integer number s there exists a real plane algebraic curve $\mathbb{R}B$ of degree sm with $sn+\frac{(s-1)(s-2)}{2}$ nests of total depth $sd+m\frac{(s-1)(s-2)}{2}$ and a line in $\mathbb{R}P^2$ intersecting $\mathbb{R}B$ in sm points and intersecting each oval of the $sn+\frac{(s-1)(s-2)}{2}$ nests in not more than 2 points.*

To make the previous lemma work we need to start with some curve.

LEMMA 4. *For any m there exists a real plane nonsingular algebraic curve $\mathbb{R}A$ of degree m with a nest of depth $[\frac{m}{2}]$ and a line in $\mathbb{R}P^2$ intersecting $\mathbb{R}A$ in m points and intersecting each oval of the nest in 2 points.*

Lemma 3 and Lemma 4 allow to construct a surfaces in $\#\mathbb{C}P^2$ realizing the homology class $(m,0,\dots,0)$ given a decomposition of m into factors. Namely, this gives us the following theorem.

THEOREM 1. *If $m=p_1p_2\dots p_r$, $p_j\in\mathbb{N}$, then the homology class*

$$(m,0,\dots,0)\in H_2(\#(1+p_2\dots p_r+\sum_{j=2}^{r}\frac{(p_j-1)(p_j-2)}{2}p_{j+1}\dots p_r)\mathbb{C}P^2)$$

is realizable by a smooth orientable embedded surface of genus

$$\frac{(m-1)(m-2)}{2}-[\frac{p_1}{2}]p_2\dots p_r-m\sum_{j=2}^{r}\frac{(p_j-1)(p_j-2)}{2p_j}+1+2[\frac{p_1}{2}]-p_1.$$

REMARK 2.1. Lemma 1 and Lemma 3 produce also other series of embedded surfaces in $\#\mathbb{C}P^2$ starting from other real algebraic curves (Lemma 3 here works inductively to produce some curves of higher degree with many nests). In the statement of Theorem 1 we included only surfaces which are produced by this machine starting from the real algebraic curves given by Lemma 4, since we need real curves with high value of $d-n$ and they are rather difficult to classify (except those of Lemma 4 and those, produced from them by Lemma 3).

3. Stabilization of the minimal genus problem for surfaces in $\mathbb{C}P^2$

3.1. Minimal genus in $\mathbb{C}P^2$. Let F be a connected smooth surface embedded into $\mathbb{C}P^2$. Then $[F] = m[\mathbb{C}P^1] \in H_2(\mathbb{C}P^2)$. We equip F with such an orientation that $m \geq 0$. If F is a nonsingular holomorphic curve in $\mathbb{C}P^2$ then

$$g(F) = \frac{(m-1)(m-2)}{2},$$

where $g(F)$ is the genus of F.

Let now $m = 2k$ be an even number. Then, by the Rokhlin-Hsiang-Szczarba inequality (see [**3**], [**2**]),

$$g(F) \geq k^2 - 1.$$

Denote by $g_{\min}(m)$ the minimal possible genus of a surface realizing class $m[\mathbb{C}P^1]$. If $m = 2k$ then from the discussion above we get the following bounds for $g_{\min}(2k)$.

$$k^2 - 1 \leq g_{\min}(2k) \leq (2k-1)(k-1). \tag{1}$$

3.2. Stable minimal genus in $\mathbb{C}P^2$. We now consider a connected smooth surface F in the connected sum $\#n\mathbb{C}P^2$ of n copies of $\mathbb{C}P^2$ of homology class $(2k, 0, \dots, 0) \in H_2(\#n\mathbb{C}P^2) \approx \mathbb{Z} \oplus \dots \oplus \mathbb{Z}$. The Rokhlin-Hsiang-Szczarba inequality gives

$$g(F) \geq k^2 - n.$$

On the other hand, if $n - 1 \leq (2k-1)(k-1)$ then there exists an irreducible holomorphic curve in $\mathbb{C}P^2$ of degree 2k with $n-1$ ordinary double points and no other singularities (it is easy to construct such a curve, for example, perturbing a union of $2k$ lines in general position). The genus of such a curve is $(2k-1)(k-1)-(n-1)$. This curve produces a connected smooth surface in $\#n\mathbb{C}P^2$ of homology class $(2k, 0, \dots, 0)$ and genus $(2k-1)(k-1)-(n-1)$ after $(n-1)$ anticomplex blowups in the double points. To see that the homology class of the new surface is $(2k, 0, \dots, 0)$ we note that if we change the orientation of one locus of the curve in each double point then we get an ordinary (-1) intersection of two planes in $\mathbb{R}^4$, so anticomplex blowup will desingularize this double point and, therefore, the intersection of the proper transform of the curve with the exceptional divisor is zero.

We now define the *stable minimal genus* $g_{\text{st}}(m)$ as the minimum value of

$$g(F) + n - 1$$

over all positive integer n and surfaces $F \subset \#n\mathbb{C}P^2$ representing the homology class $(m, 0, \dots, 0)$. From the above discussions for $g_{\text{st}}(2k)$ we still have the bounds from (1):

$$k^2 - 1 \leq g_{\text{st}}(2k) \leq (2k-1)(k-1).$$

Theorem 1 allows us to improve the upper bound.

THEOREM 2.

$$g_{\mathrm{st}}(p_1 \dots p_r) \le \frac{(p_1 \dots p_r - 1)(p_1 \dots p_r - 2)}{2} - ([\frac{p_1}{2}] - 1)p_2 \dots p_r - \sum_{j=2}^{r} (p_1 \dots p_{j-1} - 1)\frac{(p_j - 1)(p_j - 2)}{2} p_{j+1} \dots p_r.$$

Theorem 2 (together with the Rokhlin-Hsiang-Szczarba inequality) gives the following inequality in the case when $m = 2k$ is even

$$k^2 - 1 \le g_{\mathrm{st}}(2k) \le \frac{3}{2}k^2 - \frac{3}{2}k. \tag{2}$$

Applying Lemma 2 to the Harnack curve and then applying Lemma 1 we get that if $k > 0$ is even (i.e. m is divisible by 4) then

$$k^2 - 1 \le g_{\mathrm{st}}(2k) \le \frac{11}{8}k^2 - \frac{3}{4}k - 1. \tag{3}$$

The inequalities (2) and (3) can be also proven topologically as it was observed by R.Gompf. The topological proof admits a useful generalization.

THEOREM 3. *If $m = 2k > 0$ is divisible by 2^s, $s > 0$ (i.e. k is divisible by 2^{s-1}) then*

$$k^2 - 1 \le g_{\mathrm{st}} \le \frac{4 + \frac{1}{2^{2s-1}}}{3}k^2 - \frac{3}{2^s}k + 1 - s.$$

This theorem produces the most efficient upper bound for g_{st} in this paper. It can be generalized to other factorizations of m, but it would give a weaker bound (also Theorem 2 would produce a better bound for divisors of m greater than 3).

3.3. Small values of degree of homology class in $\mathbb{C}P^2$. In the following table we summarize what we know of the minimal genus $g(m)$ and the stable minimal genus $g_{\mathrm{st}}(m)$ for values of m up to 16 (the inequalities for the minimal genus come from Rokhlin-Hsiang-Szczarba inequalities and, in the case when $m \equiv 2 \pmod 4$, from the Donaldson theorem, which produces inequality $g(4r + 2) \ge (2r + 1)^2 + 1)$).

$g(1) = 0$	$g_{\mathrm{st}}(1) = 0$	$19 \le g(9) \le 28$	$19 \le g_{\mathrm{st}}(9) \le 25$
$g(2) = 0$	$g_{\mathrm{st}}(2) = 0$	$26 \le g(10) \le 36$	$24 \le g_{\mathrm{st}}(10) \le 30$
$g(3) = 1$	$g_{\mathrm{st}}(3) = 1$	$29 \le g(11) \le 45$	$29 \le g_{\mathrm{st}}(11) \le 39$
$g(4) = 3$	$g_{\mathrm{st}}(4) = 3$	$35 \le g(12) \le 55$	$35 \le g_{\mathrm{st}}(12) \le 44$
$5 \le g(5) \le 6$	$g_{\mathrm{st}}(5) = 5$	$42 \le g(13) \le 66$	$42 \le g_{\mathrm{st}}(13) \le 55$
$g(6) = 10$	$8 \le g_{\mathrm{st}}(6) \le 9$	$50 \le g(14) \le 78$	$48 \le g_{\mathrm{st}}(14) \le 63$
$11 \le g(7) \le 15$	$11 \le g_{\mathrm{st}}(7) \le 13$	$54 \le g(15) \le 91$	$54 \le g_{\mathrm{st}}(15) \le 76$
$15 \le g(8) \le 21$	$15 \le g_{\mathrm{st}}(8) \le 18$	$63 \le g(16) \le 105$	$63 \le g_{\mathrm{st}}(16) \le 80$

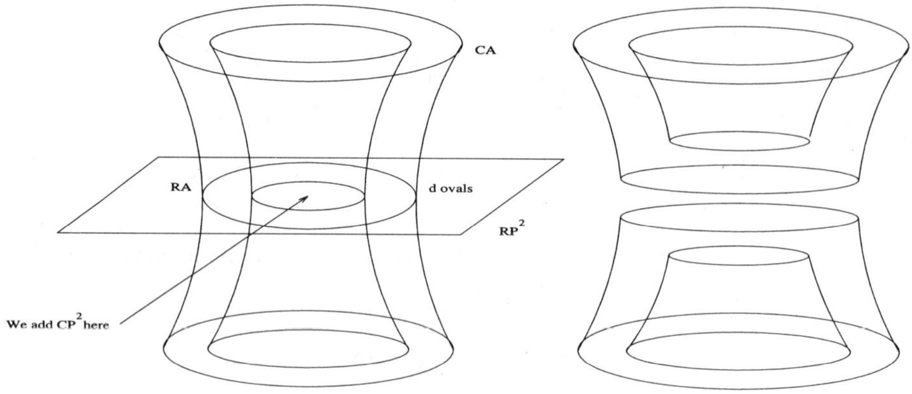

FIGURE 1. Embedded surgery

4. Proof of Lemma 1

We start with the complexification $\mathbb{C}A \subset \mathbb{C}P^2$ of the real curve $\mathbb{R}A$, a surface in $\mathbb{C}P^2$ of genus $\frac{(m-1)(m-2)}{2}$ realizing homology class $m[\mathbb{C}P^1] \in H_2(\mathbb{C}P^2)$. For each oval C of the n nests of $\mathbb{R}A$ we consider the interior disk $D \subset \mathbb{R}P^2$ such that $\partial D = C$. The disk D is the membrane for the surface $\mathbb{C}A \subset \mathbb{C}P^2$, i.e. $\partial D \subset \mathbb{C}A$. The easy computation using the isomorphism of the tangent and normal bundles of D given by the multiplication by i shows that the index of the membrane D is (-1) (i.e. the self-intersection of D suitable with the framing over ∂D given by $\mathbb{C}A$ is (-1)). Furthermore, the interior disks of the ovals of the same nest overlap over the interior disk of the innermost oval and the interior disks of the ovals from disjoint nests are disjoint.

We now make a connected sum of the pair $(\mathbb{C}P^2, D)$ with the pair $(\mathbb{C}P^2, \mathbb{C}P^1)$ for each one of the n disjoint nests of $\mathbb{R}A$, where D is the interior disk of the innermost oval of the nest. This adds 1 to the indices of the interior disk membranes of the n nests and, therefore, the new indices are equal to 0. Furthermore, after the connected sum we can make the interior disk membranes of the same nest disjoint by a small perturbation.

To finish the proof we just make the embedded surgery along the resulting disjoint 0-membranes (see fig. 1). The embedded surgery decreases the genus of the surface if the result of the surgery is connected. Surface $\mathbb{C}A - \mathbb{R}A$ contains at most two components. Therefore, surface is still connected if at least one component of $\mathbb{R}A$ is left unsurgered. Therefore, by the embedded surgery we can decrease the genus by $d-1$ in the case **a)** and by d in the case **b)**.

5. Construction of real algebraic curves

In this section we prove Lemma 2, Lemma 3 and Lemma 4. To construct the curve for Lemma 4 we take $[\frac{m}{2}]$ concentric circles (curves of degree 2) and, in the case when m is odd, an additional line disjoint from the circles (see fig. 2). This

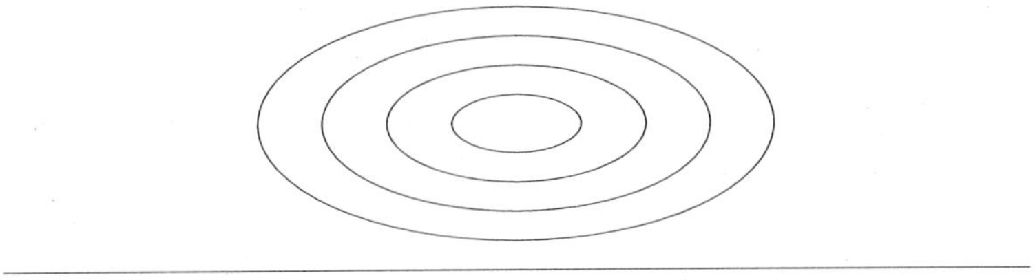

FIGURE 2. Curve for Lemma 4

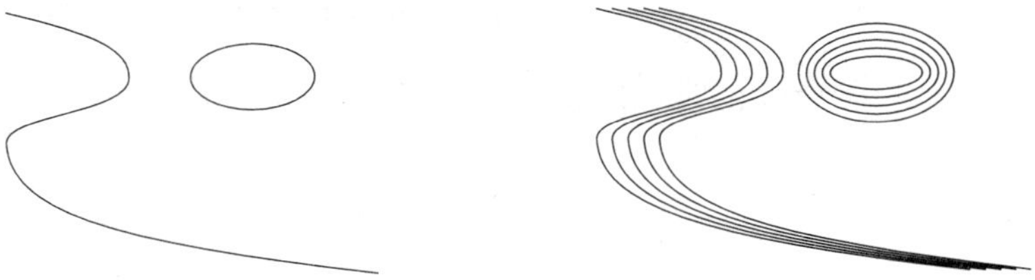

FIGURE 3. Proof of Lemma 2

is a reducible curve of degree m with a nest of depth $[\frac{m}{2}]$. All of the singular points of this curve are imaginary. A small perturbation produces a nonsingular curve with the same arrangement of the real part in $\mathbb{R}P^2$.

Let us now prove Lemma 2. Let $\{(x_0 : x_1 : x_2) \in \mathbb{R}P^2 | f(x_0, x_1, x_2) = 0\}$ be the nonsingular curve $\mathbb{R}A$ with n disjoint nests of total depth d. Then, for a small $\epsilon > 0$, equation

$$f(x_0,x_1,x_2)(f^2(x_0,x_1,x_2)-\epsilon(x_0^{2m}+x_1^{2m}))\dots(f^2(x_0,x_1,x_2)-\frac{s-1}{2}\epsilon(x_0^{2m}+x_1^{2m}))$$

if s is odd or

$$(f^2(x_0,x_1,x_2)-\epsilon(x_0^{2m}+x_1^{2m}))\dots(f^2(x_0,x_1,x_2)-\frac{s}{2}\epsilon(x_0^{2m}+x_1^{2m}))$$

if s is even, produces a reducible curve of degree sm. All singular points of this curve are imaginary and its real part may be obtained from $\mathbb{R}A$ by replacing each component of $\mathbb{R}A$ with s parallel copies of this component (note that we actually replace an oval with s ovals and a component isotopic to $\mathbb{R}P^1$ with $[\frac{s}{2}]$ ovals, and, in the case when s is odd, in addition, with a "core" component isotopic to $\mathbb{R}P^1$), see fig. 3. A small perturbation of this curve produces a nonsingular real curve of degree sm with n nests of total depth sd.

To prove Lemma 3 we use a trick of B.Chevalier [**1**]. We start from a Harnack curve of degree s with scheme

$$< \frac{3k^2-3k}{2} \sqcup 1 < \frac{k^2-3k+2}{2} >> \quad \text{(see fig. 4)}$$

in Viro's notation [**4**] with a nonempty oval intersecting the line $x_2 = 0$ in $2k$ points for even $s = 2k$ and from a Harnack curve with scheme

$$< j \sqcup k(2k-1) > \quad \text{(see fig. 5)}$$

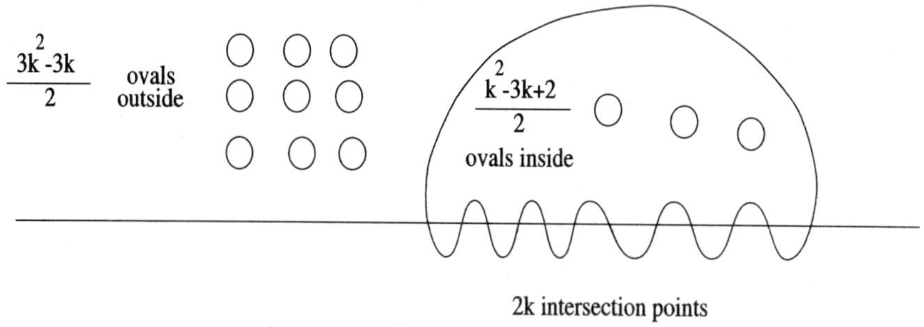

FIGURE 4. Harnack curve of degree $2k$

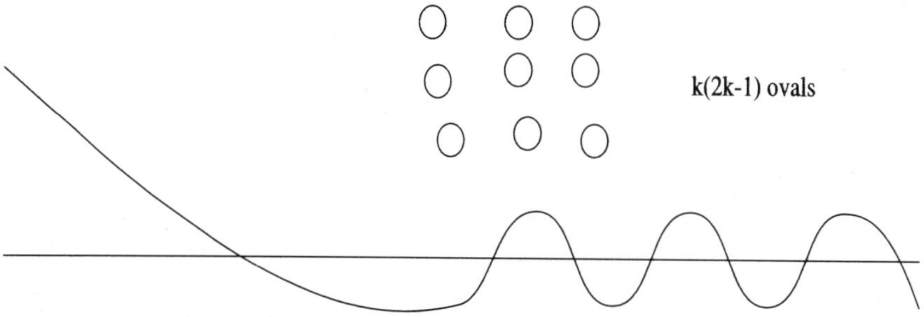

FIGURE 5. Harnack curve of degree $2k+1$

in Viro's notation [**4**] with a nonempty oval intersecting the line $x_2 = 0$ in $2k+1$ points for odd $s = 2k+1$ (for the construction see e.g. [**5**]). Let f be the homogeneous polynomial determining this Harnack curve. Consider now the curve $\mathbb{R}A$ given by the hypothesis of Lemma 3. Let g be the homogeneous polynomial determining this Harnack curve.

By projective linear transformation, stretching the neighborhood of L, we arrange the auxiliary curve $g = 0$ so that the line $x_1 = Cx_0$ intersects $\mathbb{R}A$ in the same fashion as L if $|C| > \delta$, where $\delta > 0$ is small (with respect to the curve $f = 0$), see fig. 6. Now, by [**1**], for small value of $\epsilon > 0$, we can recover the topology of the curve

$$g(\epsilon x_0^m, \epsilon x_1 x_0^{m-1}, f(x_0, x_1, x_2)) = 0$$

(see fig. 7). This equation determines the curve of degree sm. Each oval of $f = 0$ not intersecting the line $x_1 = 0$ produces the nest of depth m and, since these ovals are disjoint for our Harnack curve, they produce disjoint nests. Each nest of $g = 0$ suitably positioned with respect to L (i.e. such that each oval of this nest intersect L not more than in 2 points) produces s disjoint nests of the same depth. After a small perturbation we get a nonsingular curve with $sn + \frac{(s-1)(s-2)}{2}$ nests of total depth $sd + m\frac{(s-1)(s-2)}{2}$.

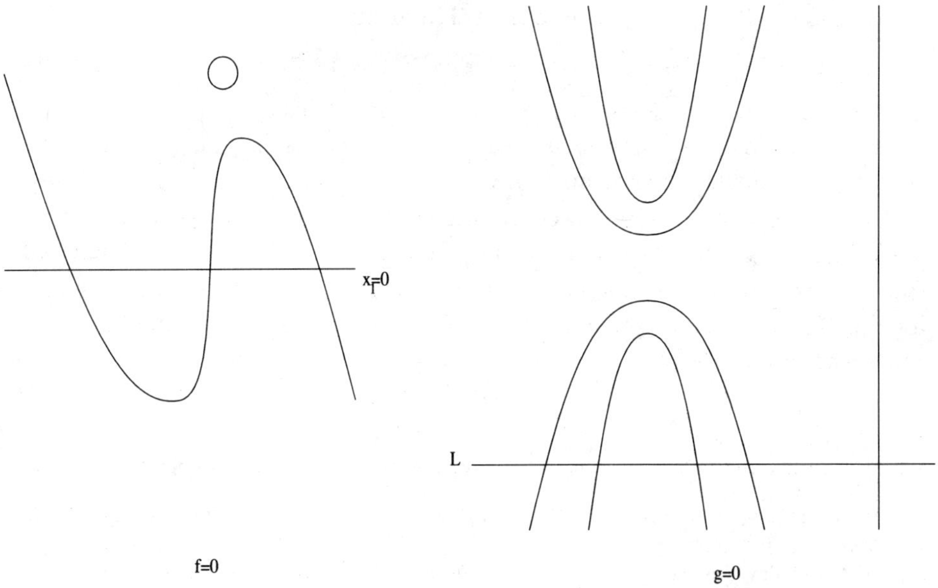

FIGURE 6. Mutual position of curves $f = 0$ and $g = 0$

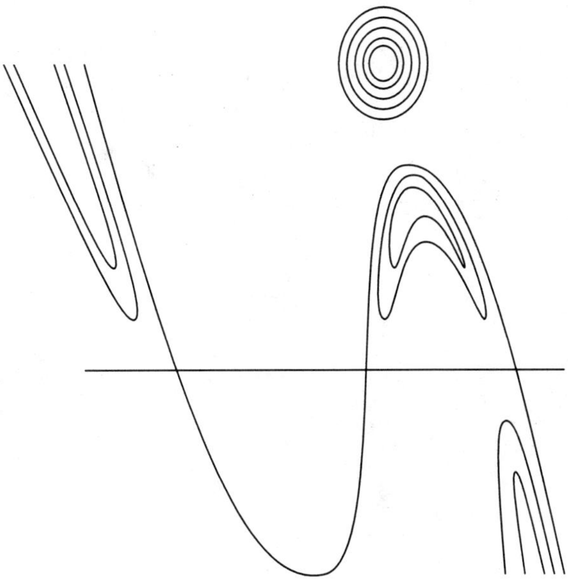

FIGURE 7. Resulting curve of the Chevallier trick

6. Proof of Theorem 3

The homology class $\frac{k}{2^{s-1}}[\mathbb{C}P^1] \in H_2(\mathbb{C}P^2)$ is realizable by a generic immersed sphere with $\frac{(\frac{k}{2^{s-1}}-1)(\frac{k}{2^{s-1}}-2)}{2}$ double points (this sphere is provided e.g. by a generic rational curve of degree $\frac{k}{2^{s-1}}$). Like in subsection 3.2 this produces an embedded sphere in $\#(1+\frac{k}{2^{s-1}})\mathbb{C}P^2$ realising the homology class $(\frac{k}{2^{s-1}}, 0, \ldots, 0)$. The self-intersection number of this sphere is $\frac{k^2}{2^{2s-2}}$, therefore, class $(\frac{k}{2^{s-2}}, 0, \ldots, 0)$ is representable in $\#(1+\frac{k}{2^{s-1}})\mathbb{C}P^2$ by a generic immersed sphere with $\frac{k^2}{2^{2s-2}} - 1$ double points. To finish the proof of the theorem we proceed by induction adding a copy of $\mathbb{C}P^2$ for each double point of the immersed sphere at each stage.

REFERENCES

1. B. Chevallier, *Courbes maximales et courbes non réduites*, C. R. Acad. Sc. Paris **301(2)** (1985), 35-40.
2. W. Hsiang, R. Szczarba, *On embedded surfaces in four-manifolds*, Proc. Sympos. Pure Math. **22** (1970), 97-103.
3. V. A. Rokhlin, *Two-dimensional submanifolds of four-dimensional manifolds*, Funktsional. Anal. i Prilozhen., **5(1)** (1971), 48-60.
4. O. Ya. Viro, *The progress in topology of real algebraic varieties over last six years*, Uspehi Mat. Nauk, **41(3)** (1986), 45-67.
5. O. Ya. Viro, *Real algebraic plane curves: constructions with controlled topology*, Leningrad Math. J., **1(5)** (1990), 1059-1134

INSTITUTE FOR ADVANCED STUDY, OLDEN LANE, PRINCETON NJ 08540 USA

Current address: Max-Plank-Institut für Mathematik, Gottfried-Claren-Strasse 26, 53225 Bonn, Germany

E-mail address: mihalkin@mpim-bonn.mpg.de

Contemporary Mathen.
Volume **182**, 1995

On Spaces of Holomorphic Maps from Two Copies of the Riemann Sphere to Complex Grassmannians

JOHN W. HAVLICEK

ABSTRACT. This paper studies the structure of the space $\mathrm{Hol}_*(\mathbb{P}^1 \times \mathbb{P}^1, \mathbb{G})$ of based holomorphic maps from $\mathbb{P}^1 \times \mathbb{P}^1$ to $\mathbb{G} = \mathbb{G}_{2,s}$, the complex Grassmannian of 2-planes in $\mathbb{C}^{s+2}$. Basepoint in $\mathbb{P}^1 \times \mathbb{P}^1$ is taken as the wedge point of the 2-skeleton $\mathbb{P}^1 \vee \mathbb{P}^1 \subset \mathbb{P}^1 \times \mathbb{P}^1$. Restriction to the 2-skeleton gives the map $\rho : \mathrm{Hol}_*(\mathbb{P}^1 \times \mathbb{P}^1, \mathbb{G}) \to (\mathrm{Rat}(\mathbb{G}))^2$, where we have written $\mathrm{Rat}(\mathbb{G}) = \mathrm{Hol}_*(\mathbb{P}^1, \mathbb{G})$. Our main result gives the structure of the fiber $\rho^{-1}(f,g)$ when f is generic of degree less than s and g is generic of degree one. We also give a complete analysis of $\rho^{-1}(f,g)$ when $s > 1$ and both f and g are of degree one.

0. Introduction

This paper studies the structure of the space $\mathrm{Hol}(\mathbb{P}^1 \times \mathbb{P}^1, \mathbb{G}_{2,s})$ of holomorphic maps from $\mathbb{P}^1 \times \mathbb{P}^1$ to the complex Grassmannian of 2-planes in $\mathbb{C}^{s+2}$. (All mapping spaces and spaces of sections are given the compact-open topology.) The restriction to Grassmannians of 2-planes is for simplicity, and the approach of this paper should be applicable to Grassmannians of higher dimensional planes. The essential complexities of $\mathrm{Hol}(\mathbb{P}^1 \times \mathbb{P}^1, \mathbb{G}_{2,s})$ appear when $s \geq 2$.

One reason for studying this space is that it represents geometry in "transcendence degree two," at least insofar as the domain of the mapping space is two-dimensional. Considerable work has been done on geometry in transcendence degree one, and in particular spaces of (based) holomorphic maps from $\mathbb{P}^1$ to projective spaces, Grassmannian manifolds, and flag manifolds have been fairly well understood. See, for example, **[S, K, CCMM, CS, Gr, Gu, MM1,**

1991 *Mathematics Subject Classification.* Primary 58D15; Secondary 32L05, 32H99.

The author was supported in part by a grant from the ARCS Foundation.

This paper is in final form and no versio: f it will be submitted for publication elsewhere.

MM2, MM3]. The analysis has centered around the standard local data for a function of one complex variable: isolated zeros and poles with multiplicity, residues, and so forth. A central theme in the work has been to understand the holomorphic mapping spaces as finite-dimensional approximations to the continuous mapping spaces into which they naturally include and to see how that approximation "stabilizes" as the topological degree (appropriately defined) of the maps becomes large. In contrast, geometry in transcendence degree two is little understood, and the local data are considerably more complicated.

Among compact complex surfaces, $\mathbb{P}^1 \times \mathbb{P}^1$ is simple and has the advantage that adjointness allows the identification

$$\mathrm{Hol}(\mathbb{P}^1 \times \mathbb{P}^1, Y) \cong \mathrm{Hol}(\mathbb{P}^1, \mathrm{Hol}(\mathbb{P}^1, Y)) \tag{0.1}$$

for Y a complex manifold. This identification will be central to the present work. From the cofiber sequence

$$\mathbb{P}^1 \vee \mathbb{P}^1 \to \mathbb{P}^1 \times \mathbb{P}^1 \to S^4,$$

we obtain the fibration of based continuous mapping spaces

$$\Omega^4 Y \to \mathrm{Map}_*(\mathbb{P}^1 \times \mathbb{P}^1, Y) \to (\Omega^2 Y)^2.$$

(The wedge point of $\mathbb{P}^1 \vee \mathbb{P}^1$ is taken as basepoint in both $\mathbb{P}^1 \vee \mathbb{P}^1$ and $\mathbb{P}^1 \times \mathbb{P}^1$.) The associated holomorphic mapping spaces include naturally into the total space and base space of this fibration:

$$\begin{array}{ccccc} & & \mathrm{Hol}_*(\mathbb{P}^1 \times \mathbb{P}^1, Y) & \xrightarrow{\rho} & (\mathrm{Rat}(Y))^2 \\ & & \downarrow & & \downarrow \\ \Omega^4 Y & \longrightarrow & \mathrm{Map}_*(\mathbb{P}^1 \times \mathbb{P}^1, Y) & \longrightarrow & (\Omega^2 Y)^2 . \end{array} \tag{0.2}$$

Here we are using the notation $\mathrm{Rat}(Y)$, which is common in the literature, to denote the space of based holomorphic maps from $\mathbb{P}^1$ to Y. The restriction map ρ is not typically a fibration. A first step toward understanding the structure of $\mathrm{Hol}_*(\mathbb{P}^1 \times \mathbb{P}^1, Y)$ is therefore analyzing the fiber of ρ over various pairs of maps $(f, g) \in (\mathrm{Rat}(Y))^2$. The main result of this paper gives the structure of the fiber $\rho^{-1}(f, g)$ when $Y = \mathbb{G}_{2,s}$, f is generic with $\deg f < s$, and g is a generic map of degree one.

The choice of the target $Y = \mathbb{G}_{2,s}$ of the holomorphic maps requires some motivation. Indeed, taking the target to be a projective space is simpler, and the results of this paper have analogues for the case $Y = \mathbb{P}^n$. When $Y = \mathbb{G}_{2,s}$, though, there is a relationship between components of certain of the fibers $\rho^{-1}(f, g)$ and the moduli space $\mathcal{M}_c$ of $SU(2)$-instantons on the four-sphere with charge c. Work of Atiyah [**A**] and Donaldson [**D**] identifies $\mathcal{M}_c$ with isomorphism classes of rank-two holomorphic vector bundles E on $\mathbb{P}^1 \times \mathbb{P}^1$ whose second Chern class is c and which are trivial on $\mathbb{P}^1 \vee \mathbb{P}^1$ with a fixed trivialization there. Let η denote the tautological 2-plane bundle on $\mathbb{G}_{2,s}$ and let η^* be its dual. The bundle E typically does not have enough holomorphic sections to be induced from η^* by a holomorphic map to $\mathbb{G}_{2,s}$, but it can be transformed into a bundle which is induced in this fashion by tensoring with an appropriate holomorphic line bundle L. The map φ which induces $E \otimes L$ is then fixed on $\mathbb{P}^1 \vee \mathbb{P}^1$ as a pair of maps (f, g) which are determined by the original trivialization of E on $\mathbb{P}^1 \vee \mathbb{P}^1$ and the choice of L. Thus, φ lies in $\rho_c^{-1}(f, g)$, the part of the fiber consisting of maps which induce the second Chern class c on $\mathbb{P}^1 \times \mathbb{P}^1$.

The main result of this paper is

THEOREM A. *Let $s > k$ and let $f \in \mathrm{Rat}_k(\mathbb{G}_{2,s})$ be chosen in a suitable generic neighborhood. Also let $0 \leq m \leq n$ be the unique integers such that $f^*\eta \cong \mathcal{O}(-m) \oplus \mathcal{O}(-n)$, where η is the tautological 2-plane bundle over $\mathbb{G}_{2,s}$. (It follows that $m + n = k$.) Then for generic g in $\mathrm{Rat}_1(\mathbb{G}_{2,s})$,*

$$\rho^{-1}(f, g) = \coprod_{c=n}^{k} \rho_c^{-1}(f, g),$$

and there are homeomorphisms

$$\rho_c^{-1}(f, g) \cong \mathrm{Rat}_{c-n} \times \mathrm{Rat}_{k-c}(\mathbb{P}^{s-1}) \times \mathbb{C}^{2n}.$$

Here, $\mathrm{Rat}_i(\mathbb{P}^j) \subset \mathrm{Rat}(\mathbb{P}^j)$ denotes the component of maps inducing $c_1 = i$, and $\mathrm{Rat}_i = \mathrm{Rat}_i(\mathbb{P}^1)$.

REMARK 0.3. The spaces $\mathrm{Rat}_k(\mathbb{P}^n)$ are well understood now from work begun by Segal [**S**] and culminating with work by F. Cohen, R. Cohen, Mann, and Milgram [**CCMM**] and work by R. Cohen and Shimamoto [**CS**]. The stable homotopy type and homology are given in [**CCMM**], while [**CS**] gives the homotopy type when $n > 1$. The spaces $\mathrm{Rat}_k(\mathbb{G}_{2,s})$ have been studied by Kirwan [**K**] and Mann and Milgram [**MM1, MM2**].

When f and g are both of degree one, we give a complete analysis of the fibers. By post-composing with an automorphism of $\mathbb{G}_{2,s}$, we can assume f has the explicit form

$$f(z : w) = \begin{pmatrix} 1 & 0 & 0 & 0 & \cdots & 0 \\ 0 & z & w & 0 & \cdots & 0 \end{pmatrix}. \tag{0.4}$$

Notice that $f \in \mathrm{Rat}_1(\mathbb{G}_{2,1})$ (i.e., the image of f lies in $\mathbb{G}_{2,1} \subset \mathbb{G}_{2,s}$). We observe in Example 1.5(a) that $\mathrm{Rat}_1(\mathbb{G}_{2,s})$ is covered by two open affine neighborhoods, $\mathcal{U}_{1,0}$ and $\mathcal{U}_{0,1}$, and the fiber $\rho^{-1}(f, g)$ depends on whether g lies in the neighborhood $\mathcal{U}_{1,0}$. The result is

THEOREM B. *Let f have the form (0.4) and assume $s > 1$.*

(a) *Let $g \notin \mathrm{Rat}_1(\mathbb{G}_{2,1})$. If $g \in \mathcal{U}_{1,0}$, then $\rho^{-1}(f, g) = \rho_1^{-1}(f, g) \cong \mathbb{C}^2$, while if $g \notin \mathcal{U}_{1,0}$, then $\rho^{-1}(f, g) = \rho_0^{-1}(f, g) \cong \mathrm{Rat}_1(\mathbb{P}^{s-1}) \times \mathbb{C}$.*

(b) *Let $g \in \mathrm{Rat}_1(\mathbb{G}_{2,1})$. Then $\rho_2^{-1}(f, g) \cong \mathbb{C}^2 \times \mathbb{C}^*$, and $\rho_1^{-1}(f, g)$ is empty. If $g \notin \mathcal{U}_{1,0}$, then $\rho_0^{-1}(f, g) \cong \mathrm{Rat}_1(\mathbb{P}^{s-1}) \times \mathbb{C}$, and if $g \in \mathcal{U}_{1,0}$, then $\rho_0^{-1}(f, g)$ is empty.*

In order to prove the theorems, we use the adjointness identification (0.1) to replace $\rho^{-1}(f,g)$ with the equivalent space of based holomorphic sections of a holomorphic fiber bundle over $\mathbb{P}^1$. There is a holomorphic bundle

$$\mathrm{Rat}(\mathbb{G}_{2,s}) \to \mathrm{Hol}(\mathbb{G}_{2,s}) \xrightarrow{e} \mathbb{G}_{2,s},$$

where $\mathrm{Hol}(Y) = \mathrm{Hol}(\mathbb{P}^1, Y)$ and e is evaluation at a point. Thus, a map f in $\mathrm{Rat}_k(\mathbb{G}_{2,s})$ induces a bundle $E(f) \to \mathbb{P}^1$ with fiber $\mathrm{Rat}(\mathbb{G}_{2,s})$:

$$\begin{array}{ccc} E(f) & \longrightarrow & \mathrm{Hol}(\mathbb{G}_{2,s}) \\ \downarrow & & \downarrow e \\ \mathbb{P}^1 & \xrightarrow{f} & \mathbb{G}_{2,s}\ . \end{array}$$

Using adjointness, we can identify a map $\varphi : \mathbb{P}^1 \times \mathbb{P}^1 \to \mathbb{G}_{2,s}$ which satisfies

$$\varphi|_{\mathbb{P}^1 \times (1:0)} = f$$

with a holomorphic section s of $E(f)$. If $\rho(\varphi) = (f,g)$, then the section s is based according to g. Thus,

$$\rho^{-1}(f,g) \cong \Gamma_g E(f),$$

where $\Gamma_g E(f)$ denotes the space of holomorphic sections of $E(f)$ that are based according to g. The space $\Gamma_g E(f)$ can be understood by proceeding in analogy with spaces of sections of holomorphic line bundles over $\mathbb{P}^1$. Using the manifold structure of the fiber, one finds normal forms for the sections over trivializing neighborhoods for the bundle $E(f)$ and checks the constraint conditions for patching them together over the intersections.

ACKNOWLEDGEMENTS. Much of the work of this paper was carried out in the author's Ph.D. thesis [**H1**] under the patient direction of R. J. Milgram and with partial support from the ARCS Foundation. The main result of [**H1**] is a variation of Theorem A in which f is chosen in an explicit way. The author also wishes to acknowledge many useful conversations with D. Gomprecht while preparing the present manuscript.

1. Preliminaries

This section gives background material on $\mathrm{Hol}(\mathbb{G}_{r,s})$ and preliminary observations about $\mathrm{Hol}(\mathbb{P}^1 \times \mathbb{P}^1, \mathbb{G}_{r,s})$. Most of the material in this section appears either implicitly or explicitly in the literature. See, for example, [**S, CCMM, MM1**]. Proofs of many of the elementary facts are given in [**H1**]. Throughout this paper, we shall abbreviate $\mathbb{G} = \mathbb{G}_{r,s}$ or $\mathbb{G} = \mathbb{G}_{2,s}$ according to the context.

1.1. Generalities. It is well-known (GAGA, [**Ser**]) that any holomorphic map $V \to W$ of smooth projective varieties is rational (i.e., given by rational functions in affine neighborhoods). In studying the structure of the mapping space $\mathrm{Hol}(V, W)$, it is desirable to pass from the local description of a map to a global (or nearly global) representation in which the constraints to extending over all of V can be more easily seen.

For example, a map f in $\mathrm{Hol}(\mathbb{P}^n, \mathbb{P}^m)$ $(m \geq n)$ is given locally (i.e., in terms of inhomogeneous coordinates) by rational functions, but it is easy to check that by converting to homogeneous coordinates f can be expressed globally as $(p_0 : \cdots : p_m)$, where the p_j are homogeneous polynomials in $n+1$ variables, all of the same degree, and with no common zeros. The common degree of the polynomials is the "degree" of f and gives the integral cohomology map induced by f in dimension two. The mapping space separates according to degree. If f is of degree k, then there are $N = \binom{n+k}{k}(m+1)$ coefficients in the polynomials $p_0, \ldots, p_m$, and so

$$\mathrm{Hol}_k(\mathbb{P}^n, \mathbb{P}^m) = \mathbb{P}^{N-1} - R, \tag{1.1}$$

where R is the resultant variety corresponding to tuples $(p_0, \ldots, p_m)$ in which $p_0, \ldots, p_m$ share a common zero.

If $V = \mathbb{P}^n$ and if $W \subseteq \mathbb{P}^m$, then

$$\mathrm{Hol}(\mathbb{P}^n, W) \subseteq \mathrm{Hol}(\mathbb{P}^n, \mathbb{P}^m),$$

so $\mathrm{Hol}(\mathbb{P}^n, W)$ separates (at least) according to degree. The polynomials which cut W from $\mathbb{P}^m$ give rise to polynomials cutting $\mathrm{Hol}_k(\mathbb{P}^n, W)$ from $\mathrm{Hol}_k(\mathbb{P}^n, \mathbb{P}^m)$, and in view of (1.1) it follows that $\mathrm{Hol}_k(\mathbb{P}^n, W)$ is a quasiprojective variety. $\mathrm{Hol}_k(\mathbb{P}^n, W)$ need not be smooth; for example, if $n \geq 2$ then $\mathrm{Hol}_k(\mathbb{P}^n, W)$ may separate further according to the cohomology map in higher dimensions. If $x_0 \in \mathbb{P}^n$ and $w_0 \in W$ are basepoints, then imposing the basing condition $f(x_0) = w_0$ on the maps amounts to intersecting $\mathrm{Hol}_k(\mathbb{P}^n, W)$ with a smooth subvariety of codimension m in $\mathbb{P}^{N-1}$. The result is that the based space $\mathrm{Rat}_k(\mathbb{P}^n, W)$ is again a quasiprojective variety. The choice of the basepoint x_0 is immaterial. If W is a symmetric space $W \cong G/F$, where G acts on W by biholomorphisms and F is the stabilizer of w_0, then the homeomorphism type of $\mathrm{Rat}_k(\mathbb{P}^n, W)$ does not depend on w_0. In fact, the fibration

$$\mathrm{Rat}_k(\mathbb{P}^n, W) \to \mathrm{Hol}_k(\mathbb{P}^n, W) \xrightarrow{e} W,$$

where e is evaluation at x_0, is the fiber bundle associated to the principal bundle $F \to G \to W$ by the post-composition action of F on $\mathrm{Rat}_k(\mathbb{P}^n, W)$:

$$\mathrm{Hol}_k(\mathbb{P}^n, W) \cong \mathrm{Rat}_k(\mathbb{P}^n, W) \times_F G. \tag{1.2}$$

1.2. Polynomial representations of maps in Hol($\mathbb{G}$). The case of present interest is $W = \mathbb{G} = \mathbb{G}_{r,s}$. The space $\mathrm{Rat}_k(\mathbb{G})$ has been studied extensively in [**MM1**], to which the reader is referred for further details. It is well-known (see [**C, ByH, MM1**]) that $\mathrm{Rat}_k(\mathbb{G})$ is a connected, noncompact complex manifold of dimension $k(r+s)$. $\mathbb{G}$ may be viewed as the quotient GL_{r+s}/F, where F is the subgroup of (r,s) block lower triangular matrices, that is matrices in the block form

$$\begin{pmatrix} F_{11} & 0 \\ F_{21} & F_{22} \end{pmatrix}, \quad F_{11} \in GL_r, \ F_{22} \in GL_s. \tag{1.3}$$

The principal bundle $F \to GL_{r+s} \to \mathbb{G}$ is holomorphic, and the action of F on $\mathrm{Rat}_k(\mathbb{G})$ is holomorphic, so it follows that

$$\mathrm{Hol}_k \cong \mathrm{Rat}_k(\mathbb{G}) \times_F GL_{r+s} \tag{1.4}$$

is a connected, noncompact complex manifold of dimension $k(r+s)+rs$.

The manifold structure of $\mathrm{Rat}_k(\mathbb{G})$ can be seen by putting the maps into appropriate "normal forms". For our purposes, a normal form for a map is a matrix of polynomials that represents the map throughout an affine line and which satisfies extra normalizing conditions. The normal forms given by Mann and Milgram lead to a stratification of $\mathrm{Rat}_k(\mathbb{G})$ in which the top-dimensional stratum is a manifold chart. A slight modification of their treatment yields an atlas of manifold charts for $\mathrm{Rat}_k(\mathbb{G})$. These normal forms and charts are reviewed in §1.3 below for the case $r = 2$.

For the remainder of this subsection, we take $r = 2$ and abbreviate $\mathbb{G} = \mathbb{G}_{2,s}$. (See [**MM1**] or [**H1**] for a general treatment.) Let $\mathbb{V}$ denote the Stiefel manifold of $2 \times (s+2)$ matrices of rank two. Then $\mathbb{G} = \mathbb{V}/GL_2$, where GL_2 acts by left multiplication. Manifold charts for $\mathbb{G}$ are given as follows. If $I = (i,j)$ is a multi-index with $1 \le i < j \le s+2$, then let U_I denote the subset of matrices in $\mathbb{V}$ for which the minor determined by columns i and j (in order) is the 2×2 identity matrix (the other matrix entries therefore being free). Then $U_I \cong \mathbb{C}^{2s}$ is biholomorphic to its image under the projection $\mathbb{V} \to \mathbb{G}$ and gives a chart for $\mathbb{G}$. Basepoint in $\mathbb{G}$ is taken as

$$w_0 = (\, id_2 \quad 0 \,),$$

the "origin" of U_{12}. We also fix the charts $U_0 = \{(1:w)\}$ and $U_1 = \{(z:1)\}$ in $\mathbb{P}^1$. Basepoint in $\mathbb{P}^1$ is taken as $(1:0)$.

Rationalness ensures that in the affine neighborhoods U_1 and U_{12}, a map $f \in \mathrm{Hol}(\mathbb{G})$ is represented locally by a matrix $B = (\, id_2 \quad B_0 \,)$, where B_0 is a $2 \times (s+2)$ matrix of rational functions in the coordinate z on U_1. The entries of B_0 can be taken as quotients of coprime polynomials. If f is a based map, then $B_0 \to 0$ as $z \to \infty$, so each nonzero entry of B_0 has denominator of degree strictly greater than the numerator. B represents f throughout the complement in U_1 of the singularities of B_0.

In order to obtain a more global representation of f, we clear the denominators of the fractions by multiplying each row of B by the lcm of the denominators in that row. The result is a matrix B_1 with polynomial entries and such that the entries in a single row have gcd $= 1$. Then there is a matrix $M \in GL_{s+2}(\mathbb{C}[z])$ and a matrix $N \in SL_2(\mathbb{C}[z])$ such that

$$NB_1M = \begin{pmatrix} 1 & 0 & \cdots & 0 \\ 0 & b_2 & \cdots & b_{s+2} \end{pmatrix}.$$

(See [**H1**], for example.) Let B_2 be obtained by dividing the second row of NB_1M by $\gcd(b_2, \ldots, b_{s+2})$, and let $A = B_2M^{-1}$. Then A has polynomial entries, and the determinants of the 2×2 minors of A are coprime. Thus, the matrix A represents f throughout the affine line U_1. Using the coprime condition on the determinants of the minors, it is not difficult to check that A is unique up to left multiplication by an element of $GL_2(\mathbb{C}[z])$.

Recall that the Plücker embedding of $\mathbb{G}$ in $\mathbb{P}^m$ ($m = \binom{s+2}{2} - 1$) can be lifted to a map $\mathbb{V} \to \mathbb{C}^{m+1}$ which sends a matrix to the tuple of determinants of its 2×2 minors in a fixed order. It follows that the degree of f is the maximum degree of the determinants of the 2×2 minors of A. If f is based, then the determinant of the minor consisting of the first two columns must have strictly maximal rank. We write

$$A = (\,P \quad Q\,),$$

where P is 2×2. Normal forms for based maps are obtained by using the left action of $GL_2(\mathbb{C}[z])$ to put the matrix P into a normalized form.

1.3. The Mann–Milgram stratification and coordinate charts. Let P be an $n \times n$ matrix with polynomial entries. We say that P is in *standard upper triangular* (*s.u.t.*) *form* if P is upper triangular with monic diagonal entries and in each column of P the diagonal entry is of strictly maximal degree. It is not difficult to verify that given any $n \times n$ polynomial matrix P with nonzero determinant, there is a unique matrix $M \in GL_n(\mathbb{C}[z])$ such that MP is in s.u.t. form; furthermore the s.u.t. matrix MP is uniquely determined by P. In [**MM1**], Mann and Milgram normalize the matrices representing maps in $\mathrm{Rat}(\mathbb{G})$ by requiring P to be in s.u.t. form. Thus, they require

$$A = (\,P \quad Q\,) = \begin{pmatrix} p_1 & r & q_1 \\ 0 & p_2 & q_2 \end{pmatrix},$$

where p_1, p_2 are monic polynomials; r is a polynomial with $\deg r < \deg p_2$; and q_1, q_2 are s-tuples of polynomials. With $k_i = \deg p_i$, the degree of the map represented by A is $k = k_1 + k_2$. The basing condition requires that $P^{-1}Q \to 0$ as $z \to \infty$. Since

$$P^{-1}Q = \begin{pmatrix} p_1^{-1}p_2^{-1}(p_2q_1 - rq_2) \\ p_2^{-1}q_2 \end{pmatrix},$$

it follows that the polynomials in q_2 must have degree strictly less than k_2 and the polynomials in $p_2q_1 - rq_2$ must have degree strictly less than k. The polynomials in q_1 need not have degree strictly less than k_1, but the coefficients which appear in degrees k_1 and above are determined by P and q_2 and are therefore not free.

The strata of $\mathrm{Rat}_k(\mathbb{G})$ are indexed by pairs of nonnegative integers $K = (k_1, k_2)$ satisfying $k_1 + k_2 = k$. The stratum X_K consists of the maps whose normal form $(P \quad Q)$ has $\deg p_i = k_i$. By counting the free coefficients (i.e., the coefficients in P, q_2, and the coefficients in q_1 which are not determined by P and q_2), we see that X_K is smooth of dimension $k(s+1) + k_2$. (X_K is not contractible; recall that the determinants of the minors of $(P \quad Q)$ are constrained to be coprime.) Notice that the stratum of maximum dimension is $X_{0,k}$, which is an open, dense subset of $\mathrm{Rat}_k(\mathbb{G})$. If $r \geq s$, then $X_{r,k-r}$ is in the closure of $X_{s,k-s}$.

The uniqueness of s.u.t. form guarantees that the strata X_K for different pairs K do not intersect. In order to obtain coordinate charts (by which we mean affine coordinate neighborhoods, not necessarily contractible), the normalization of the matrix P must be relaxed in order to allow overlaps. For a pair K, we say that a 2×2 matrix

$$P = \begin{pmatrix} p_1 & r_2 \\ r_1 & p_2 \end{pmatrix}$$

is in *K-standard form* if the diagonal entry p_i is monic of degree k_i and the off-diagonal entries satisfy $\deg r_i < k_i$. We say that P *admits* K-standard form if there is a matrix $M \in GL_2(\mathbb{C}[z])$ such that MP is in K-standard form. It is straightforward to check that if $\det P$ has degree k, then P admits K-standard form for some K satisfying $k_1 + k_2 = k$, and in fact generic P admit K-standard form for all such K. Futhermore, if P admits K-standard form, then the matrix M and the K-standard form MP are uniquely determined by P.

To obtain an affine coordinate chart for $\mathrm{Rat}_k(\mathbb{G})$, we normalize the minor P to have K-standard form ($k_1 + k_2 = k$). Each choice of K gives a chart $\mathcal{U}_K$, and a map in $\mathcal{U}_K$ is represented by a unique matrix

$$A = (P \quad Q) = \begin{pmatrix} p_1 & r_2 & q_1 \\ r_1 & p_2 & q_2 \end{pmatrix}.$$

The basing constraint $P^{-1}Q \to 0$ as $z \to \infty$ again determines free and bound coefficients in Q. Note that

$$P^{-1}Q = p_1^{-1}p_2^{-1} \begin{pmatrix} p_2q_1 - r_2q_2 \\ p_1q_2 - r_1q_1 \end{pmatrix}.$$

Thus, each of $p_2q_1 - r_2q_2$ and $p_1q_2 - r_1q_1$ must have degree strictly less than k. Since $\deg r_i < k_i = \deg p_i$, it follows that the coefficients in q_i of degree $\geq k_i$ are determined (by induction involving both q_1 and q_2) from P and the free coefficients of Q (i.e., the coefficients of degree less than k_i in each q_i). The coefficients in P and the free coefficients in Q are affine coordinates, and we see that $\mathcal{U}_K$ has the correct dimension $k(s+2)$. $X_K \subset \mathcal{U}_K$ as the maps whose matrix P is upper triangular, and the coeffients of r_1 which must be set to zero account for its codimension k_1. Of course, $X_{0,k} = \mathcal{U}_{0,k}$.

EXAMPLES 1.5.

(a) $\mathrm{Rat}_1(\mathbb{G})$. There are only two choices for K, $(1,0)$ and $(0,1)$.

$$\mathcal{U}_{1,0} = \left\{ \begin{pmatrix} z+a & 0 & c \\ b & 1 & 0 \end{pmatrix} \,\middle|\, a, b \in \mathbb{C},\ c \in \mathbb{C}^s - \{0\} \right\}.$$

Similarly,

$$\mathcal{U}_{0,1} = \left\{ \begin{pmatrix} 1 & \beta & 0 \\ 0 & z+\alpha & \gamma \end{pmatrix} \,\middle|\, \alpha, \beta \in \mathbb{C},\ \gamma \in \mathbb{C}^s - \{0\} \right\}.$$

Uniqueness of s.u.t. form ensures that overlap occurs precisely when $b \neq 0$ and $\beta \neq 0$, in which case the transitions are

$$\alpha = a, \ \beta = 1/b, \ \gamma = -bc.$$

Thus,

$$\mathrm{Rat}_1(\mathbb{G}) \cong \mathbb{C} \times \left[(\mathbb{C}^2 - \{0\}) \times_{\mathbb{C}^*} (\mathbb{C}^s - \{0\}) \right],$$

where the action of $\mathbb{C}^*$ identifies

$$(b_0\lambda, b_1\lambda) \times c \sim (b_0, b_1) \times \lambda c$$

for $\lambda \in \mathbb{C}^*$. Put another way, $\mathrm{Rat}_1(\mathbb{G})$ is the total space of a fiber bundle

$$\mathbb{C} \times (\mathbb{C}^s - \{0\}) \to \mathrm{Rat}_1(\mathbb{G}) \to \mathbb{P}^1.$$

(b) $\mathrm{Rat}_2(\mathbb{G})$. K is one of $(0,2)$, $(1,1)$, and $(2,0)$. A map in $\mathcal{U}_{0,2}$ has the form

$$\begin{pmatrix} 1 & b_1 z + b_0 & b_1 c_1 \\ 0 & z^2 + a_1 z + a_0 & c_1 z + c_0 \end{pmatrix},$$

where $c_0, c_1 \in \mathbb{C}^s$. The determinants of the 2×2 minors must be coprime. A map in $\mathcal{U}_{1,1}$ has the form

$$\begin{pmatrix} z+\alpha_1 & \beta_1 & \gamma_1 \\ \beta_2 & z+\alpha_2 & \gamma_2 \end{pmatrix},$$

where $\gamma_1, \gamma_2 \in \mathbb{C}^s$. Again, the determinants of the 2×2 minors must be coprime. The overlap occurs when $b_1 \neq 0$ and $\beta_2 \neq 0$, in which case the transitions are

$$\begin{cases} a_1 = \alpha_1 + \alpha_2 \\ a_0 = \alpha_1\alpha_2 - \beta_1\beta_2 \\ b_1 = 1/\beta_2 \\ b_0 = \alpha_2/\beta_2 \\ c_1 = \gamma_2 \\ c_0 = \alpha_1\gamma_2 - \beta_2\gamma_1, \end{cases}$$

and

$$\begin{cases} \alpha_1 = a_1 - b_0/b_1 \\ \alpha_2 = b_0/b_1 \\ \beta_1 = -a_0 b_1 + a_1 b_0 - b_0^2/b_1 \\ \beta_2 = 1/b_1 \\ \gamma_1 = a_1 b_1 c_1 - b_0 c_1 - b_1 c_0 \\ \gamma_2 = c_1. \end{cases}$$

From the symmetry of the space, one obtains analogous transitions from $\mathcal{U}_{1,1}$ to $\mathcal{U}_{2,0}$. The more complicated transitions from $\mathcal{U}_{0,2}$ to $\mathcal{U}_{2,0}$ can be derived directly or deduced from the preceding transitions.

1.4. Preliminaries on $\mathrm{Hol}(\mathbb{P}^1 \times \mathbb{P}^1, \mathbb{G})$**.** Notice that $\mathrm{Hol}(\mathbb{P}^1 \times \mathbb{P}^1, \mathbb{P}^m)$ separates according the image of the maps in cohomology. This image is specified by a pair of integers $(k, l) \in H^2(\mathbb{P}^1 \times \mathbb{P}^1)$, and adjointness identifies

$$\mathrm{Hol}_{(k,l)}(\mathbb{P}^1 \times \mathbb{P}^1, \mathbb{P}^m) \cong \mathrm{Hol}_k(\mathrm{Hol}_l(\mathbb{P}^m)).$$

Since $\mathrm{Hol}_l(\mathbb{P}^m)$ is the complement in a projective space $\mathbb{P}^N$ of a variety, it follows that $\mathrm{Hol}_k(\mathrm{Hol}_l(\mathbb{P}^m))$ is also the complement in a projective space $\mathbb{P}^M$ of a variety. (The polynomials that cut the complement of $\mathrm{Hol}_l(\mathbb{P}^m)$ from $\mathbb{P}^N$ give rise to polynomials that cut the complement of $\mathrm{Hol}_k(\mathrm{Hol}_l(\mathbb{P}^m))$ from $\mathbb{P}^M$.) Restricting the image to $\mathbb{G} = \mathbb{G}_{r,s} \subset \mathbb{P}^m$ is accomplished by intersecting with a variety, so $\mathrm{Hol}_{(k,l)}(\mathbb{P}^1 \times \mathbb{P}^1, \mathbb{G})$ is quasiprojective.

$\mathrm{Hol}_{(k,l)}(\mathbb{P}^1 \times \mathbb{P}^1, \mathbb{G})$ separates further according to the cohomology map in H^4, i.e., according to the Chern class c_2. It is not known in general whether c_2 determines a component of $\mathrm{Hol}_{(k,l)}(\mathbb{P}^1 \times \mathbb{P}^1, \mathbb{G})$, nor whether the components of $\mathrm{Hol}_{(k,l)}(\mathbb{P}^1 \times \mathbb{P}^1, \mathbb{G})$ are smooth.

2. Meromorphic sections of line bundles

It will be useful in studying sections of the bundle $E(f)$ to have fixed notation and reviewed basic facts about meromorphic sections of holomorphic line bundles over $\mathbb{P}^1$. The affine neighborhoods U_i of $\mathbb{P}^1$ are as in §1. The affine coordinate on U_0 is w, the affine coordinate on U_1 is z, and $z = w^{-1}$ in $U_0 \cap U_1$.

2.1. Definitions. Recall that a holomorphic line bundle over $\mathbb{P}^n$ is classified by its first Chern class. We write $\mathcal{O}(k)$ for the bundle with $c_1 = k$, and we abbreviate $\mathcal{O}(0) = \mathcal{O}$. Over $\mathbb{P}^1$, $\mathcal{O}(k)$ can be trivialized on each affine neighborhood U_i in such a way that a section of $\mathcal{O}(k)$ is determined by a pair of maps $s_i : U_i \to \mathbb{C}$ transforming according to

$$s_1(z) = z^k s_0(z^{-1}). \tag{2.1}$$

The section $s = (s_0, s_1)$ is *holomorphic* or *meromorphic* according as the component functions s_i are holomorphic or meromorphic. Of course, a meromorphic section of $\mathcal{O}(k)$ is not properly a section of the bundle $\mathcal{O}(k)$, but is a holomorphic section of the associated $\mathbb{P}^1$-bundle, the Hirzebruch surface

$$\mathbb{F}_k = \mathbb{P}[\mathcal{O}(k+r) \oplus \mathcal{O}(r)]. \tag{2.2}$$

Here, $\mathbb{P}E$ is the projectivization of the complex bundle E. $\mathbb{P}E$ is obtained as $E_0/\mathbb{C}^*$, where E_0 is the complement in E of the zero section and $\mathbb{C}^*$ acts by scalar multiplication in the fibers. Different choices of r give isomorphic projectivized bundles in (2.2).

We use $\Gamma\mathbb{F}_k$ to denote the space of meromorphic sections of $\mathcal{O}(k)$ (i.e., holomorphic sections of $\mathbb{F}_k$). The points $(0:1)$ and $(1:0)$ are well-defined in every fiber of $\mathbb{F}_k$, and they give rise to two corresponding "constant" sections in $\Gamma\mathbb{F}_k$, one with value $(0:1)$ and the other with value $(1:0)$. We may embed $\mathcal{O}(k) \to \mathbb{F}_k$ as the complement of the section with value $(1:0)$. This convention leads us to call $(0:1)$ the *zero section* and $(1:0)$ the *infinity section* in $\Gamma\mathbb{F}_k$. We then recover the notion of zero and pole of a section $s \in \Gamma\mathbb{F}_k$ as an intersection of the image of s with the zero or infinity section. We also identify the holomorphic sections of $\mathcal{O}(k)$ with the sections in $\Gamma\mathbb{F}_k$ whose images do not meet the infinity section.

2.2. Separation in $\Gamma\mathbb{F}_k$. From the transition (2.1) it follows that a section $s \in \Gamma\mathbb{F}_k$ which is not the zero or infinity section must satisfy

$$\#\{\text{zeros of } s\} - \#\{\text{poles of } s\} = k. \tag{2.3}$$

If $k > 0$, then the zero section can be deformed into the holomorphic sections and has self-intersection k; the infinity section is holomorphically rigid and has self intersection $-k$. If $k < 0$, then the zero section is holomorphically rigid with self intersection k, while the infinity section can be deformed into the sections which do not meet the zero section and has self-intersection $-k$. When $k = 0$, both the zero section and the infinity section can be deformed into the constant sections, and each has self intersection zero. The symmetry here illustrates the fact that the involution of $\mathbb{P}^1$ sending $(x:y) \mapsto (y:x)$ induces a homeomorphism

$$\Gamma\mathbb{F}_k \cong \Gamma\mathbb{F}_{-k} \tag{2.4}$$

which exchanges zeros and poles.

The section space $\Gamma\mathbb{F}_k$ separates according to the number of zeros or poles of a section. We choose to index the components by the number of poles, and we write $\Gamma^r\mathbb{F}_k$ to denote the component whose sections have a total of r poles. The convention is that $r \geq \max(0, -k)$. When rigid, the zero section or infinity section is a single point component. According to the discussion above, we have for $k > 0$

$$\Gamma\mathbb{F}_k = \{(1:0)\} \amalg \coprod_{r \geq 0} \Gamma^r\mathbb{F}_k \; ; \tag{2.5}$$

for $k < 0$

$$\Gamma\mathbb{F}_k = \{(0:1)\} \coprod \coprod_{r \geq -k} \Gamma^r\mathbb{F}_k \; ; \tag{2.6}$$

and for $k = 0$

$$\Gamma\mathbb{F}_0 = \coprod_{r \geq 0} \Gamma^r\mathbb{F}_0 \, , \tag{2.7}$$

where $\Gamma^0\mathbb{F}_0 = \mathbb{P}^1$ denotes the component of constant sections. When $k > 0$, $\Gamma\mathcal{O}(k) = \Gamma^0\mathbb{F}_k$; when $k = 0$, $\Gamma\mathcal{O} = \mathbb{C} \subset \mathbb{P}^1 = \Gamma^0\mathbb{F}_0$, the complement of the infinity section; finally, when $k < 0$, $\Gamma\mathcal{O}(k)$ is empty.

2.3. The structure of $\Gamma^r\mathbb{F}_k$ and $\Gamma^r_*\mathbb{F}_k$. A section $s \in \Gamma^r\mathbb{F}_k$ can be lifted to a holomorphic section of $[\mathcal{O}(r+k) \oplus \mathcal{O}(r)]_0$, and the lift is unique up to multiplication by elements of $\mathbb{C}^*$. For $m, n \geq 0$, the holomorphic section space $\Gamma[\mathcal{O}(m) \oplus \mathcal{O}(n)]_0$ can be identified with a space of polynomial pairs as follows. Let $\mathcal{P}(m,n)$ denote the space of coprime pairs (p,q) of polynomials in $\mathbb{C}[z]$ satisfying $\deg p \leq m$ and $\deg q \leq n$, but such that equality holds for at least one. Then a section $s = (s_0, s_1)$ of $[\mathcal{O}(r+k) \oplus \mathcal{O}(r)]_0$ may be identified with the pair (p,q) such that $s_1(z) = (p(z), q(z))$. The coprime condition ensures that s does not vanish in U_1, while the condition on degrees ensures that s_1 extends to give a section of $\mathcal{O}(m) \oplus \mathcal{O}(n)$ which does not vanish at the point $w = 0$ of U_0. (Notice that we could equally well choose to identify $s_0(w) = (p(w), q(w))$, which amounts to reversing the order of the coefficients of the polynomials associated to s.) When $\max(m,n) > 0$,

$$\mathcal{P}(m,n) = \mathbb{C}^{m+n+2} - R,$$

where R is the affine hypersurface cut out by the resultant of a general polynomial of degree m and one of degree n. Since the resultant is homogeneous, we can factor by the action of $\mathbb{C}^*$ to obtain

$$\Gamma^r\mathbb{F}_k = \mathcal{P}(r+k, r)/\mathbb{C}^* = \mathbb{P}^{2r+k+1} - \mathbb{P}R, \tag{2.8}$$

which exhibits $\Gamma^r\mathbb{F}_k$ as the complement of a projective hypersurface and as the base of a holomorphic principal $\mathbb{C}^*$-bundle. In case $r = k = 0$, $\mathcal{P}(0,0) = \mathbb{C}^2 - \{(0,0)\}$, so we recover $\Gamma^0\mathbb{F}_0 = \mathbb{P}^1$. If $r = 0$ and $k > 0$, then $\mathbb{P}R = \mathbb{P}^k$ is cut out by $q = 0$, so

$$\Gamma\mathcal{O}(k) = \Gamma^0\mathbb{F}_k = \mathbb{C}^{k+1}, \tag{2.9}$$

as expected. Similarly, if $r = -k$ and $k < 0$, then

$$\Gamma^{-k}\mathbb{F}_k = \mathbb{C}^{-k+1}. \tag{2.10}$$

Apart from these special cases, the topology of $\Gamma^r\mathbb{F}_k$ can be complicated and has not been given in general. In view of the homeomorphism (2.4), we restrict attention to $k \geq 0$. The structures of $\mathcal{P}(k+1, 1)$ and $\Gamma^1\mathbb{F}_k$ are not difficult to determine, although they are not needed for the present work. More substantially, $\Gamma^r\mathbb{F}_0$ is the same as Hol_r, the space of degree r holomorphic self-maps of $\mathbb{P}^1$. Hol_r and the subset $\mathrm{Rat}_r \subset \mathrm{Hol}_r$ consisting of based maps have been studied extensively (see [**S, CCMM, CS, T, H2**]). For the present work, the structure of the based section spaces $\Gamma^r_*\mathbb{F}_k$ and $\Gamma_*[\mathcal{O}(r+k)\oplus\mathcal{O}(r)]_0$ are more important. Here we think of a section s in terms of component functions s_0, s_1, and basing is accomplished by specifying the value that one of the functions s_i is to take at a certain point.

For generic choices of basing, the based section space $\Gamma^r_*\mathbb{F}_k$ is homotopy equivalent to Rat_r. This fact, a consequence of long division, was observed by Segal [**S**], and we review it here. We can base over the point $w = 0$ by specifying that a section $s \in \Gamma^r_*\mathbb{F}_k$ satisfy $s_0(0) = (\lambda_0 : \lambda_1) \in \mathbb{P}^1$. Let S be a lift of s to $\Gamma[\mathcal{O}(r+k)\oplus\mathcal{O}(r)]_0$, and let S correspond to the polynomial pair $(p,q) \in \mathcal{P}(r+k,r)$ according to $S_1(z) = (p(z), q(z))$. Let p_0 be the coefficient of z^{r+k} in p and let q_0 be the coefficient of z^r in q. Then $(p_0 : q_0) = (\lambda_0 : \lambda_1)$. If

$$\lambda_1 \neq 0,$$

then $\deg q = r$. The lift S is then uniquely determined by requiring q to be monic, and $p_0 = \lambda_0/\lambda_1$ is determined by the basing. We divide q into p, obtaining

$$p = q\alpha + \beta,$$

where α is of degree at most k and β is of degree $< r$. The coefficient of z^k in α is p_0, and so it is fixed by the basing, but the other coefficients are unconstrained. The polynomial β must satisfy $\gcd(q,\beta) = 1$, and thus the polynomial pair (q,β) is equivalent to an element of Rat_r. Then we have the biholomorphism

$$\Gamma^r_*\mathbb{F}_k \cong \mathrm{Rat}_r \times \mathbb{C}^k \tag{2.11}$$

by sending s to $(q,\beta)\times(\alpha - p_0 z^k)$. Of course, the same proof shows that if the basing value (λ_0, λ_1) satisfies $\lambda_1 \neq 0$, then

$$\Gamma_*[\mathcal{O}(r+k)\oplus\mathcal{O}(r)]_0 \cong \mathrm{Rat}_r \times \mathbb{C}^k. \tag{2.12}$$

For the proof of Theorem B, we need complete results in the case of $\Gamma^1_*\mathbb{F}_1$ and $\Gamma_*[\mathcal{O}(2)\oplus\mathcal{O}(1)]_0$.

LEMMA 2.13. *For any choice of basing, the section spaces $\Gamma^1_*\mathbb{F}_1$ and $\Gamma_*[\mathcal{O}(2)\oplus\mathcal{O}(1)]_0$ are both biholomorphic to $\mathbb{C}^2\times\mathbb{C}^*$.*

PROOF. The notation is as above. Since $\mathrm{Rat}_1 \cong \mathbb{C}\times\mathbb{C}^*$, we need only determine what happens when the basing is taken as $q_0 = 0$. In this case, $p_0 \neq 0$, so the lift S is uniquely determined by requiring that p be monic. The constant term of q can be any element of $\mathbb{C}^*$, and the remaining coefficients of p can be taken freely in $\mathbb{C}^2$. □

2.4. Sections of higher rank bundles. The Birkhoff-Grothendieck Theorem ensures that a rank n holomorphic vector bundle E over $\mathbb{P}^1$ splits holomorphically as a direct sum of line bundles

$$E \cong \bigoplus_{i=1}^{n} \mathcal{O}(k_i), \tag{2.14}$$

from which

$$\Gamma E \cong \prod_{i=1}^{n} \Gamma \mathcal{O}(k_i). \tag{2.15}$$

If an index $k_i < 0$, then a holomorphic section must be identically zero in the corresponding coordinate. Thus, we lose no generality by assuming that all $k_i \geq 0$. For our purposes, the holomorphic sections which do not meet the zero section are of primary interest. As in §2.3, we have

$$\Gamma E_0 \cong \mathcal{P}(k_1, \dots, k_n),$$

where $\mathcal{P}(k_1, \dots, k_n)$ denotes the space of n-tuples of polynomials $(p_1, \dots, p_n)$ satisfying $\gcd(p_i) = 1$ and $\deg p_i \leq k_i$ with at least one equality. The analogue of (2.12) is the following lemma, which can also be attributed to Segal.

LEMMA 2.16. *Assume that $k_i = k + t_i$, where $k, t_i \geq 0$ and $t_1 = 0$. If the basing requires the coefficient of z^k in p_1 to be nonzero (a generic condition), then*

$$\Gamma_* E_0 \cong \mathrm{Rat}_k(\mathbb{P}^{n-1}) \times \mathbb{C}^N,$$

where $N = \sum t_i$.

PROOF. The basing determines the coefficient of z^{k+t_i} in p_i. Long division gives

$$p_i = p_1 \alpha_i + \beta_i,$$

where $\deg \alpha_i \leq t_i$ and $\deg \beta_i < k$. The coefficient of z^{t_i} in α_i is determined, while the lower coefficients are unconstrained. The polynomials $p_1, \beta_2, \dots, \beta_n$ are coprime, and so $(p_1, \beta_2, \dots, \beta_n)$ is equivalent to an element of $\mathrm{Rat}_k(\mathbb{P}^{n-1})$. The biholomorphism is now assembled as in the case $n = 2$. □

3. The Bundle $E(f)$

Basepoints in $\mathbb{P}^1$ and $\mathbb{G} = \mathbb{G}_{2,s}$ are as in §1. As in (1.4), the evaluation map at basepoint $e : \mathrm{Hol}(\mathbb{G}) \to \mathbb{G}$ is the projection in the holomorphic fiber bundle

$$\mathrm{Rat}(\mathbb{G}) \to \mathrm{Hol}(\mathbb{G}) \xrightarrow{e} \mathbb{G}$$

associated to the holomorphic principal bundle

$$F \to GL_{s+2} \xrightarrow{e'} \mathbb{G}$$

by the post-composition action of F on $\mathrm{Rat}(\mathbb{G})$. If $f \in \mathrm{Hol}(\mathbb{G})$, then f induces a holomorphic fiber bundle $E(f)$ and an underlying holomorphic principal F-bundle $P(f)$:

$$\begin{array}{ccccccc} E(f) & \xrightarrow{\bar{f}} & \mathrm{Hol}(\mathbb{G}) & & P(f) & \xrightarrow{f'} & GL_{s+2} \\ \downarrow & & \downarrow e & , & \downarrow & & \downarrow e' \\ \mathbb{P}^1 & \xrightarrow{f} & \mathbb{G} & & \mathbb{P}^1 & \xrightarrow{f} & \mathbb{G} \end{array} .$$

Note that $E(f)$ separates into components according to the separation of $\mathrm{Hol}(\mathbb{G})$ by degree. We write

$$E_k(f) = f^*\mathrm{Hol}_k(\mathbb{G}) = \mathrm{Rat}_k(\mathbb{G}) \times_F P(f). \tag{3.1}$$

A holomorphic section s of $E(f)$ corresponds to a map $\varphi \in \mathrm{Hol}(\mathbb{P}^1 \times \mathbb{P}^1, \mathbb{G})$ according to

$$\varphi(x \times z) = \bar{f}(sx)(z). \tag{3.2}$$

Since $e\bar{f}s = f$, we see that

$$\varphi|_{\mathbb{P}^1 \times (1:0)} = f,$$

and so it follows that

$$\rho^{-1}(f, -) \cong \Gamma E(f).$$

Notice also that $s(1:0)$ determines

$$\bar{f}(s(1:0)) = \varphi|_{(1:0)\times\mathbb{P}^1}. \tag{3.3}$$

The degree k of $\bar{f}(s(1:0))$ determines the component $E_k(f)$ in which the image of s lies. If $\Gamma_g E(f)$ denotes the holomorphic sections of $E(f)$ which are based according to

$$\bar{f}(s(1:0)) = g \tag{3.4}$$

(and which are therefore sections of $E_{\deg g}(f)$), then

$$\rho^{-1}(f, g) \cong \Gamma_g E(f).$$

3.1. The principal bundle $P(f)$. The bundle $P(f)$ trivializes over the affine lines $U_i \subset \mathbb{P}^1$. We give a proof which generates an explicit transition function for the bundle.

A trivialization of $P(f)$ over U_i amounts to a lift $B_i : U_i \to GL_{s+2}$ such that $e'B_i = f|_{U_i}$. Note that the projection e' factors as

$$GL_{s+2} \xrightarrow{\varepsilon} \mathbb{V} \xrightarrow{\pi} \mathbb{G}$$

where ε sends a matrix to its first two rows. A polynomial representation $A_i = (P \quad Q)$ for f on U_i as in §1.2 is a lift of f over U_i to the Stiefel manifold $\mathbb{V}$: $\pi A_i = f|_{U_i}$. Thus we only need to see that A_i can be taken as the first two rows of a matrix $B_i \in GL_{s+2}(\mathbb{C}[x])$, where x is the affine coordinate on U_i. That such

B_i can always be found is a consequence of the following proposition from linear algebra (see [**H1**], for example, for a proof):

PROPOSITION 3.5. *Let A be an $r \times (r+s)$ matrix with entries in a PID R. There is a matrix $B \in GL_{r+s}(R)$ whose first r rows are A if and only if the the determinants of the $r \times r$ minors of A have* $\gcd = 1$.

An explicit transition function $T_{01} : U_0 \cap U_1 \to F$ for $P(f)$, and hence for $E(f)$, is then

$$T_{01} = B_0 B_1^{-1}. \tag{3.6}$$

A little care should be taken, since we are thinking of B_i as a matrix of polynomials in the affine coordinate on U_i. Using the conventions of §1, we can write

$$T_{01}(z) = B_0(z^{-1})B_1(z)^{-1} \quad \text{or} \quad T_{01}(w) = B_0(w)B_1(w^{-1})^{-1}. \tag{3.7}$$

A section of $E(f)$ is given by a pair of maps $\sigma_i : U_i \to \mathrm{Rat}(\mathbb{G})$ satisfying

$$\sigma_1 = \sigma_0 T_{01} \tag{3.8}$$

in $U_0 \cap U_1$.

3.2. Isomorphism classes of holomorphic principal F-bundles. We now turn to isomorphism classes of holomorphic principal F-bundles. The discussion is easily carried out in slightly more generality than we need. We take $F \subset GL_{r+s}$ as the (r,s) block lower triangular matrices (as in (1.3)), and we consider holomorphic bundles over a complex manifold X. The isomorphism classes of holomorphic principal F-bundles are in one-to-one correspondence with isomorphism classes of holomorphic (r,s) flag bundles. By an (r,s) flag bundle, we mean a pair (V,W), where V is a rank $r+s$ vector bundle and $W \subset V$ is a rank r subbundle; an isomorphism of flag bundles $\varphi : (V,W) \to (V',W')$ is a vector bundle isomorphism $\varphi : V \to V'$ which restricts to give an isomorphism of subbundles $\varphi : W \to W'$.

The correspondence is obtained by associating to a principal bundle P (with F acting on the left) the rank $r+s$ bundle V obtained from the standard right action of F on $\mathbb{C}^{r+s}$:

$$V = \mathbb{C}^{r+s} \times_F P.$$

The subspace $\mathbb{C}^r \subset \mathbb{C}^{r+s}$ in which the last s coordinates are zero is invariant under the action of F, and so

$$W = \mathbb{C}^r \times_F P$$

is a rank r subbundle of V. Thus, P gives rise to the flag bundle (V,W). On the other hand, given a flag bundle (V,W), we may take the transition functions for V to have values in F, with the upper left block giving the transitions for W, and these transition functions give an associated principal F-bundle P. It is straightforward to check that, up to isomorphism, these associations are inverses and give the aforementioned bijection.

Clearly, the isomorphism classes of flag bundles can be divided according to the isomorphism types of V and W. Suppose that (V_0, W_0) is a particular flag bundle and let $\mathcal{F} = \mathcal{F}(V_0, W_0)$ denote the isomorphism classes of flag bundles (V, W) in which $V \cong V_0$ and $W \cong W_0$. Also let $\mathrm{Hom}_0(W_0, V_0)$ denote the space of nondegenerate (i.e., rank preserving) bundle maps $W_0 \to V_0$.

PROPOSITION 3.9. *The flag bundles of the form (V_0, W) with $W \cong W_0$ are in one-to-one correspondence with*

$$\mathrm{Hom}_0(W_0, V_0)\Big/\mathrm{Aut}\ W_0.$$

$\mathcal{F}$ is in one-to-one correspondence with

$$\mathrm{Hom}_0(W_0, V_0)\Big/\mathrm{Aut}\ W_0 \times \mathrm{Aut}\ V_0.$$

Here, Aut W_0 *acts by pre-composition and* Aut V_0 *acts by post-composition.*

PROOF. Any element of $\mathcal{F}$ is represented by a flag bundle of the form (V_0, W) together with an isomorphism $W_0 \to W$. These data are equivalent to a nondegenerate bundle map $\varphi : W_0 \to V_0$ (W is just the image of φ). Let $\psi \in \mathrm{Hom}_0(W_0, V_0)$ be another bundle map. The images φW_0 and ψW_0 are the same precisely if ψ is obtained by pre-composing φ with an automorphism of W_0. The flag bundles $(V_0, \varphi W_0)$ and $(V_0, \psi W_0)$ are isomorphic precisely if ψ is obtained by post-composing φ with an automorphism of V_0. □

LEMMA 3.10. *The flag bundle associated to the principal bundle*

$$F \to GL_{r+s} \xrightarrow{\pi} \mathbb{G} = \mathbb{G}_{r,s}$$

is isomorphic to $(\mathcal{O}^{r+s}, \eta)$, where $\mathcal{O}^{r+s}$ is the trivial bundle of rank $r+s$ and η is the tautological r-bundle:

$$\eta = \left\{(v,p) \in \mathbb{C}^{r+s} \times \mathbb{G} \middle|\ v \in p\right\}.$$

PROOF. Let

$$V = \mathbb{C}^{r+s} \times_F GL_{r+s},\ W = \mathbb{C}^r \times_F GL_{r+s}$$

be the associated vector bundles, where we think of the elements of $\mathbb{C}^{r+s}$ as row vectors and $\mathbb{C}^r \subset \mathbb{C}^{r+s}$ as the vectors whose last s coordinates are zero. Let $p \in \mathbb{G}$ and let $M \in \pi^{-1}(p)$ be a lift of p to GL_{r+s} (i.e., the first r rows of M form a basis for p). If N_i denotes row i of the matrix N, then we can form a nowhere zero section

$$\varphi_i : \mathbb{G} \to V$$

by

$$\varphi_i : p \mapsto \langle (M^{-1})_i, M \rangle.$$

For $f \in F$, $((fM)^{-1})_i = (M^{-1})_i f^{-1}$, so φ_i is well-defined. Furthermore, since the lifts M are nonsingular matrices, it follows that $\varphi_1, \ldots, \varphi_{r+s}$ form a global frame for V and give a global trivialization

$$\varphi : \mathbb{C}^{r+s} \times \mathbb{G} \overset{\cong}{\to} V$$

by

$$\varphi : (v, p) \mapsto \sum v_i \varphi_i(p) = \langle v M^{-1}, M \rangle.$$

Finally, $v \in p$ is equivalent to $v = wM$ for some $w \in \mathbb{C}^r \subset \mathbb{C}^{r+s}$, in which case $\varphi(v, p) = \langle w, M \rangle \in W$. It follows that $\varphi : (\mathcal{O}^{r+s}, \eta) \to (V, W)$ is an isomorphism. □

A holomorphic map $f : X \to \mathbb{G}$ thus determines the flag bundle $(\mathcal{O}^{r+s}, f^*\eta)$ on X. Conversely, a flag bundle $(\mathcal{O}^{r+s}, W)$ on X determines a holomorphic map $f : X \to \mathbb{G}$ by letting $f(x)$ be the plane spanned by the fiber $W_x \subset \mathbb{C}^{r+s}$, and it is not hard to see that $(\mathcal{O}^{r+s}, f^*\eta)$ is isomorphic to the original flag bundle $(\mathcal{O}^{r+s}, W)$. This shows that there is a one-to-one correspondence

$$\text{(3.11)} \qquad \mathrm{Hol}(X, \mathbb{G}) \leftrightarrow \bigcup_W \mathrm{Hom}_0(W, \mathcal{O}^{r+s})/\mathrm{Aut}\ W,$$

where the union is over representatives W of the isomorphism classes of r-bundles on X (which can be embedded as subbundles of $\mathcal{O}^{r+s}$). From Proposition 3.9, it follows that two holomorphic maps $f, f' \in \mathrm{Hol}(X, \mathbb{G})$ induce isomorphic flag bundles $(\mathcal{O}^{r+s}, f^*\eta)$ and $(\mathcal{O}^{r+s}, f'^*\eta)$ if and only if $f'^*\eta$ is obtained from $f^*\eta$ by an automorphism of $\mathcal{O}^{r+s}$. Of course, Aut $\mathcal{O}^{r+s} = GL_{r+s}$, although the action of the scalar matrices does not affect the flag bundles. Under the correspondence of flag bundles and holomorphic maps determined by Proposition 3.9 and (3.11), it follows that an automorphism of $\mathcal{O}^{r+s}$ which sends $f^*\eta$ to $f'^*\eta$ can be identified with an automorphism $\alpha \in \mathbb{P}GL_{r+s} = \mathrm{Aut}\ \mathbb{G}$ such that $f' = \alpha f$. Thus, we have proved

PROPOSITION 3.12. *Two maps $f, f' \in \mathrm{Hol}(X, \mathbb{G})$ induce isomorphic holomorphic principal F-bundles from $F \to GL_{r+s} \to \mathbb{G}$ if and only if $f' = \alpha f$ for some $\alpha \in \mathbb{P}GL_{r+s}$. The set of isomorphism classes of holomorphic principal F-bundles whose associated $\mathbb{C}^{r+s}$ bundle is trivial is thus in one-to-one correspondence with*

$$\mathrm{Hol}(X, \mathbb{G})/\mathbb{P}GL_{r+s} \cong \mathrm{Rat}(X, \mathbb{G})/\mathbb{P}F.$$

Returning to the case of original interest ($r = 2$, $X = \mathbb{P}^1$), we have

COROLLARY 3.13. *$P(f) \cong P(f')$ if and only if $f' = \alpha f$ for some $\alpha \in \mathbb{P}GL_{s+2}$.*

Note that for $f \in \mathrm{Hol}_k(\mathbb{G})$, $f^*\eta$ splits as

$$\text{(3.14)} \qquad f^*\eta \cong \mathcal{O}(-m) \oplus \mathcal{O}(-n),$$

where $k = -c_1 f^*\eta = m + n$. Furthermore, the fact that $f^*\eta$ is a subbundle of $\mathcal{O}^{s+2}$ forces $0 \leq m, n$, and so we can assume $0 \leq m \leq n$.

3.3. Examples. Throughout this subsection, $\mathbb{G} = \mathbb{G}_{2,s}$.

3.3.1. *f of degree one.* Then $f^*\eta \cong \mathcal{O} \oplus \mathcal{O}(-1)$. Indeed, it is easy to check using the explicit representations of Example 1.5(a) that there is an automorphism α so that $f' = \alpha f$ is represented by

$$\begin{pmatrix} 1 & 0 & 0 & 0 & \cdots & 0 \\ 0 & z & 1 & 0 & \cdots & 0 \end{pmatrix}$$

in U_1. Then f' has the global form

$$f'(z:w) = \begin{pmatrix} 1 & 0 & 0 & 0 & \cdots & 0 \\ 0 & z & w & 0 & \cdots & 0 \end{pmatrix}$$

given in (0.4). The lift B_i can clearly be taken in $(3, s-1)$ block diagonal form, with the $s-1$ block simply the identity. Thus, the transition T_{01} will also be in $(3, s-1)$ block diagonal form, with the $s-1$ block the identity. The upper left blocks of B_1 and B_0 can be taken as

$$\begin{pmatrix} 1 & 0 & 0 \\ 0 & z & 1 \\ 0 & -1 & 0 \end{pmatrix}, \quad \begin{pmatrix} 1 & 0 & 0 \\ 0 & 1 & w \\ 0 & 0 & 1 \end{pmatrix} \tag{3.15}$$

(respectively). Then the upper left block of $T_{01}(z)$ is

$$\begin{pmatrix} 1 & 0 & 0 \\ 0 & z^{-1} & 0 \\ 0 & 1 & z \end{pmatrix}. \tag{3.16}$$

Note the transition for $\mathcal{O} \oplus \mathcal{O}(-1)$ in the upper left.

3.3.2. *f generic of degree k.* We give a description of the bundle $P(f)$ for generic f under the assumption $s \geq k$. The generic choice of f lies in the top Mann–Milgram stratum $X_{0,k}$, and so is represented in U_1 by

$$\begin{pmatrix} 1 & r & t_1 & \cdots & t_s \\ 0 & p & q_1 & \cdots & q_s \end{pmatrix}, \tag{3.17}$$

where $p, r, q_i, t_i \in \mathbb{C}[z]$ (z the coordinate on U_1); p is monic of degree k; and r, q_i are of degree at most $k-1$. The coefficients of t_i are all bound (i.e., determined from p, r, q_i under the basing condition). We are free to adjust the representation (3.17) by right multiplication by elements of F without changing the isomorphism type of $P(f)$. This is the same as doing column operations (with $\mathbb{C}$ scalars) with the restriction that the first two columns cannot be added to nor exchanged with any of the last s columns.

First, working only with the polynomials q_i, we can generically arrange that $q_i = z^{k-i}$ for $i \leq k$, and $q_i = 0$ for $i > k$. (Here is where we use the assumption that $s \geq k$.) Then we can arrange that $p = z^k$, and finally we can use the first column to ensure that the constant term of r is zero. Once these normalizations have been made, the form of the polynomials t_i with bound coefficients is easily determined. The determinant $rq_i - z^k t_i$ must have degree less than k, so we see that t_i is zero for $i \geq k$. For $i < k$, we have that $rz^{k-i} - z^k t_i$ has degree less than k, so $r - z^i t_i$ has degree less than i, which says that t_i is precisely the polynomial part of $z^{-i}r$. The form of f on U_1 can thus be taken as

$$A_1 = \begin{pmatrix} 1 & r & t_1 & \cdots & t_{k-1} & 0 & 0 & \cdots & 0 \\ 0 & z^k & z^{k-1} & \cdots & z & 1 & 0 & \cdots & 0 \end{pmatrix}, \tag{3.18}$$

where r and t_i are as specified above.

The matrix A_1 is unique in the sense that if M is an element of F (or GL_{s+2}) and $A_1 M$ is again in the form (3.18), then $A_1 M = A_1$. Thus, the coefficients of r determine coordinates on a piece of the space of isomorphism classes of bundles $P(f)$. The last $s - k$ columns of zeros in A_1 contribute no essential information and correspond to a $(s-k) \times (s-k)$ identity matrix in the lower right block of T_{01}. Thus, we shall assume $s = k$ in the discussion below.

We next obtain a representation A_0 for f on U_0. The coordinate on U_0 is w, and $zw = 1$ throughout the overlap $U_0 \cap U_1$. First, replace $z = w^{-1}$ in A_1, so that r, t_i are now polynomials in w^{-1}. Multiply the second row of A_1 by w^k and then subtract r times the result from the first row, obtaining

$$A_0 = \begin{pmatrix} 1 & 0 & \tau_1 & \cdots & \tau_k \\ 0 & 1 & w & \cdots & w^k \end{pmatrix}. \tag{3.19}$$

Here, $\tau_i = t_i - w^i r$. Since t_i consists of the terms of $w^i r$ of degree ≤ 0 in w, it follows that τ_i consists of the terms of $-w^i r$ of degree > 0 in w. Thus A_0 is a suitable representation for f on U_0.

As lifts $\bar{B}_i$ of A_i to GL_{k+2}, we can take

$$\bar{B}_0 = \begin{pmatrix} 1 & 0 & \tau_1 & \tau_2 & \cdots & \tau_k \\ 0 & 1 & w & w^2 & \cdots & w^k \\ 0 & 0 & 1 & w & \cdots & w^{k-1} \\ 0 & 0 & 0 & 1 & \cdots & w^{k-2} \\ \vdots & \vdots & \vdots & \vdots & \ddots & \vdots \\ 0 & 0 & 0 & 0 & \cdots & 1 \end{pmatrix}$$

and

$$\bar{B}_1 = \begin{pmatrix} 1 & r & t_1 & t_2 & \cdots & t_{k-1} & 0 \\ 0 & z^k & z^{k-1} & z^{k-2} & \cdots & z & 1 \\ 0 & -1 & 0 & 0 & \cdots & 0 & 0 \\ 0 & -z & -1 & 0 & \cdots & 0 & 0 \\ 0 & -z^2 & -z & -1 & \cdots & 0 & 0 \\ \vdots & \vdots & \vdots & \vdots & \ddots & \vdots & \vdots \\ 0 & -z^{k-1} & -z^{k-2} & -z^{k-3} & \cdots & -1 & 0 \end{pmatrix}$$

Although these choices are not the simplest, they lead to a transition $\bar{T}_{01}$ which has a simple form. Indeed,

$$\bar{B}_1^{-1} = \begin{pmatrix} 1 & 0 & r - zt_1 & t_1 - zt_2 & \cdots & t_{k-2} - zt_{k-1} & t_{k-1} \\ 0 & 0 & -1 & 0 & \cdots & 0 & 0 \\ 0 & 0 & z & -1 & \cdots & 0 & 0 \\ 0 & 0 & 0 & z & \cdots & 0 & 0 \\ \vdots & \vdots & \vdots & \vdots & \ddots & \vdots & \vdots \\ 0 & 0 & 0 & 0 & \cdots & z & -1 \\ 0 & 1 & 0 & 0 & \cdots & 0 & z \end{pmatrix},$$

and so

$$\bar{T}_{01} = \bar{B}_0\bar{B}_1^{-1} = \begin{pmatrix} 1 & \tau & 0 & 0 & \cdots & 0 \\ 0 & z^{-k} & 0 & 0 & \cdots & 0 \\ 0 & z^{1-k} & z & 0 & \cdots & 0 \\ 0 & z^{2-k} & 0 & z & \cdots & 0 \\ \vdots & \vdots & \vdots & \vdots & \ddots & \vdots \\ 0 & 1 & 0 & 0 & \cdots & z \end{pmatrix}, \tag{3.20}$$

where we have written

$$\tau = \tau_k = -z^{-k}r.$$

If $\alpha_i \in GL_{s+2}(\mathbb{C}[x])$ (x the affine coordinate on U_i) has $(2, s)$ block lower triangular form, then α_i defines a holomorphic map $U_i \to F$, and so $B_i = \alpha_i\bar{B}_i$ may also be taken as a lift. The new transition matrix is then

$$T_{01} = \alpha_0\bar{T}_{01}\alpha_1^{-1}.$$

It will be useful for the work in §4 to choose maps α_i so that the 2×2 block in the upper left of T_{01} is diagonal. The matrix

$$A = \begin{pmatrix} 1 & \tau \\ 0 & z^{-k} \end{pmatrix}$$

in the upper left of $\bar{T}_{01}$ gives the transition for $f^*\eta$. A can be diagonalized to give the isomorphism (3.14). Thus,

$$MAN^{-1} = \begin{pmatrix} z^{-m} & 0 \\ 0 & z^{-n} \end{pmatrix}, \tag{3.21}$$

where $M \in SL_2(\mathbb{C}[w])$ and $N \in SL_2(\mathbb{C}[z])$. For generic τ, the integers m and n will be as close as possible, equal if k is even and differing by one if k is odd. The coefficients in the entries of M and N depend in a complicated fashion on the coefficients of τ. (See [**Hu**] for further discussion.)

A few observations will be useful for the computations in §4.4. (3.21) gives

$$MA = \begin{pmatrix} z^{-m} & 0 \\ 0 & z^{-n} \end{pmatrix} N \text{ and } NA^{-1} = \begin{pmatrix} z^{m} & 0 \\ 0 & z^{n} \end{pmatrix} M,$$

which show that the polynomials in the first row of each of M and N can have degree no more than m, while those in the second row of each can have degree no more than n. Furthermore, $N_{11} = z^m M_{11}$ and $N_{21} = z^n M_{21}$, which shows that

$$(N_{11}, N_{21}) \in \mathcal{P}(m,n). \tag{3.22}$$

Now let

$$\alpha_0 = \begin{pmatrix} M & 0 \\ 0 & id_s \end{pmatrix} \text{ and } \alpha_1 = \begin{pmatrix} N & 0 \\ 0 & id_s \end{pmatrix}. \tag{3.23}$$

Then we obtain the new transition matrix

$$T_{01} = \begin{pmatrix} z^{-m} & 0 & 0 & 0 & \cdots & 0 \\ 0 & z^{-n} & 0 & 0 & \cdots & 0 \\ -z^{1-k}N_{21} & z^{1-k}N_{11} & z & 0 & \cdots & 0 \\ -z^{2-k}N_{21} & z^{2-k}N_{11} & 0 & z & \cdots & 0 \\ \vdots & \vdots & \vdots & \vdots & \ddots & \vdots \\ -N_{21} & N_{11} & 0 & 0 & \cdots & z \end{pmatrix}. \tag{3.24}$$

(Here we have relied explicitly on the fact that $\det N = 1$. If $\det N \neq 1$, then N_{11} and N_{21} must be divided by $\det N$ in (3.24) and (3.26) below.) The new lift $B_0 = \alpha_0 \bar{B}_0$ is

$$\begin{pmatrix} M_{11} & M_{12} & \tau_1 M_{11} + wM_{12} & \cdots \\ M_{21} & M_{22} & \tau_1 M_{21} + wM_{22} & \cdots \\ 0 & 0 & 1 & \cdots \\ 0 & 0 & 0 & \cdots \\ \vdots & \vdots & \vdots & \\ 0 & 0 & 0 & \cdots \end{pmatrix}, \tag{3.25}$$

and $B_1 = \alpha_1 \bar{B}_1$ is

$$\begin{pmatrix} N_{11} & z^m M_{12} & t_1 N_{11} + z^{k-1} N_{12} & \cdots \\ N_{21} & z^n M_{22} & t_1 N_{21} + z^{k-1} N_{22} & \cdots \\ 0 & -1 & 0 & \cdots \\ 0 & -z & -1 & \cdots \\ 0 & -z^2 & -z & \cdots \\ \vdots & \vdots & \vdots & \\ 0 & -z^{k-1} & -z^{k-2} & \cdots \end{pmatrix}. \tag{3.26}$$

Notice that by taking $k = 1$, $r = \tau = 0$, $t_i = \tau_i = 0$, $m = 0$, $n = 1$, and $M = N = id_2$, these formulas reduce to those in the preceding example for the degree one maps.

3.3.3. *A specific example.* To illustrate the ideas of the preceding example, we carry out the computations for a specific choice of f. We take $\mathbb{G} = \mathbb{G}_{2,3}$ and select $f \in \mathrm{Rat}_3(\mathbb{G})$ to be represented in U_1 by

$$A_1 = \begin{pmatrix} 1 & z^2 + 2z & z+2 & 1 & 0 \\ 0 & z^3 & z^2 & z & 1 \end{pmatrix}.$$

Then $r = z^2 + 2z$, so $\tau = -2w^2 - w$ and

$$A_0 = \begin{pmatrix} 1 & 0 & 0 & -2w & -2w^2 - w \\ 0 & 1 & w & w^2 & w^3 \end{pmatrix}.$$

The 2×2 upper left block of $\bar{T}_{01}$ is

$$A = \begin{pmatrix} 1 & -2w^2 - w \\ 0 & z^{-3} \end{pmatrix},$$

which must be diagonalized. It is easy to check that with the special linear matrices

$$M = \begin{pmatrix} w - \frac{1}{2} & 2 \\ -w & -2 \end{pmatrix} \text{ and } N = \begin{pmatrix} -\frac{1}{2}z + 1 & \frac{1}{2} \\ -z & 1 \end{pmatrix}$$

we obtain

$$MAN^{-1} = \begin{pmatrix} w & 0 \\ 0 & w^2 \end{pmatrix} = \begin{pmatrix} z^{-1} & 0 \\ 0 & z^{-2} \end{pmatrix}.$$

Thus, $m = 1, n = 2$, $N_{11} = -\frac{1}{2}z + 1$, and $N_{21} = -z$. Also

$$B_0 = \begin{pmatrix} w - \frac{1}{2} & 2 & 2w & w & \frac{1}{2}w \\ -w & -2 & -2w & 0 & w^2 \\ 0 & 0 & 1 & w & w^2 \\ 0 & 0 & 0 & 1 & w \\ 0 & 0 & 0 & 0 & 1 \end{pmatrix},$$

$$B_1 = \begin{pmatrix} -\frac{1}{2}z + 1 & 2z & 2 & 1 & \frac{1}{2} \\ -z & -2z^2 & -2z & 0 & 1 \\ 0 & -1 & 0 & 0 & 0 \\ 0 & -z & -1 & 0 & 0 \\ 0 & -z^2 & -z & -1 & 0 \end{pmatrix},$$

and

$$T_{01} = \begin{pmatrix} z^{-1} & 0 & 0 & 0 & 0 \\ 0 & z^{-2} & 0 & 0 & 0 \\ z^{-1} & z^{-2} - \frac{1}{2}z^{-1} & z & 0 & 0 \\ 1 & z^{-1} - \frac{1}{2} & 0 & z & 0 \\ z & 1 - \frac{1}{2}z & 0 & 0 & z \end{pmatrix}.$$

4. Holomorphic sections of $E_1(f)$

For the remainder of this paper, $\mathbb{G} = \mathbb{G}_{2,s}$. We analyze the section spaces $\Gamma E_1(f)$ for f degree one and for generic f of degree k as in §3.3.2. Although these cases could be treated together, we discuss the degree one maps first and in more detail since the notation is considerably simpler.

A holomorphic section of $E(f)$ is given by a pair of holomorphic maps $s_i : U_i \to \mathrm{Rat}(\mathbb{G})$ satisfying

$$s_1(z) = s_0(z^{-1})T_{01}(z) \tag{4.1}$$

in $U_0 \cap U_1$. The section $s = (s_0, s_1)$ corresponds to a map $\varphi \in \mathrm{Hol}(\mathbb{P}^1 \times \mathbb{P}^1, \mathbb{G})$ according to (3.2), and the rationalness of φ guarantees that the maps s_i must be given by rational functions in affine neighborhoods of $\mathrm{Rat}(\mathbb{G})$. The section s is of $E_1(f)$ precisely if the maps s_i have image in $\mathrm{Rat}_1(\mathbb{G})$.

4.1. Rational maps $\mathbb{C} \to \mathrm{Rat}_1(\mathbb{G})$. In this subsection we characterize rational maps $\mathbb{C} \to \mathrm{Rat}_1(\mathbb{G})$. Recall the coordinates a, b, c and α, β, γ and the transitions

$$\alpha = a, \ \beta = 1/b, \ \gamma = -bc \tag{4.2}$$

for $\mathrm{Rat}_1(\mathbb{G})$ given in Example 1.5(a). Since c and γ take values in $\mathbb{C}^s - \{0\}$, we write $c^1, \ldots, c^s$ for the components of c and similarly for the components of γ. A map $\sigma : \mathbb{C} \to \mathrm{Rat}_1(\mathbb{G})$ is *rational* precisely if its coordinates a, b, c and α, β, γ are rational functions of the affine coordinate on $\mathbb{C}$ and are such that the triples (a, b, c) and (α, β, γ) do not both have poles at any point $x \in \mathbb{C}$.

The transitions (4.2) impose restrictions on the triples (a, b, c) and (α, β, γ) associated to a rational map σ. Of course, if the image of σ does not meet one of the charts $\mathcal{U}_{0,1}$ and $\mathcal{U}_{1,0}$, then the corresponding coordinates are undefined and the coordinates for the other chart must be polynomials. If the image of σ does not meet $\mathcal{U}_{0,1}$, then $b \equiv 0$, a and c^t are polynomials, and $\gcd(c^t) = 1$. Similarly, if the image of σ does not meet $\mathcal{U}_{1,0}$, then $\beta \equiv 0$, α and γ^t are polynomials, and $\gcd(\gamma^t) = 1$. Now assume that the image of σ meets both charts. Then we can write $b = p/q$ where p and q are coprime polynomials and $\deg q > 0$. The zeros of q are the points at which the image of σ passes out of the chart $\mathcal{U}_{1,0}$. The transition $\alpha = a$ ensures that both α and a are polynomials. If x is a zero of q, then γ^t must not have a pole at x, and so

$$\gamma^t = -bc^t$$

shows that c^t must have a zero at x of order at least equal the order of the zero of q at x. It follows that c^t must be a polynomial and $q | c^t$. Furthermore, since c has image in $\mathbb{C}^s - \{0\}$, the polynomials c^t cannot share a common zero at a point where q is not zero. Since γ must also have image in $\mathbb{C}^s - \{0\}$, the polynomials c^t cannot share common zeros in addition to those of q, i.e.

$$\gcd(c^t) = q. \tag{4.3}$$

Equivalently, we find that γ^t are polynomials with

$$\gcd(\gamma^t) = p. \tag{4.4}$$

4.2. $\Gamma E_1(f)$ for $f \in \mathrm{Rat}_1(\mathbb{G})$. Let $s \in \Gamma E_1(f)$, where $f \in \mathrm{Rat}_1(\mathbb{G})$. As in §3.3.1, we can adjust by an automorphism of $\mathbb{G}$ to put f in the explicit form (0.4). The component function s_i is rational in the affine coordinate on U_i, and so is given by coordinate functions (a_i, b_i, c_i) and/or $(\alpha_i, \beta_i, \gamma_i)$ as in §4.1. For the moment we shall assume that the coordinates (a_i, b_i, c_i) are defined. In this case,

$$s_i = \begin{pmatrix} \zeta + a_i & 0 & c_i \\ b_i & 1 & 0 \end{pmatrix}. \tag{4.5}$$

Using the form of $T_{01}(z)$ from (3.16), the transition (4.1) is

$$\begin{aligned} s_0 T_{01} &= \begin{pmatrix} \zeta + a_0 & c_0^1 & zc_0^1 & c_0^2 & \cdots & c_0^s \\ b_0 & z^{-1} & 0 & 0 & \cdots & 0 \end{pmatrix} \\ &= \begin{pmatrix} \zeta + a_0 - zb_0c_0^1 & 0 & zc_0^1 & c_0^2 & \cdots & c_0^s \\ zb_0 & 1 & 0 & 0 & \cdots & 0 \end{pmatrix} \\ &= s_1 \ , \end{aligned}$$

which gives

$$\text{(4.6)} \qquad \begin{cases} a_1 = a_0 - zb_0c_0^1 \\ b_1 = zb_0 \\ c_1^1 = zc_0^1 \\ c_1^t = c_0^t & t > 1 \ . \end{cases}$$

The coordinates (a_i, b_i, c_i) are undefined precisely when $\beta_i \equiv 0$. In this case,

$$\text{(4.7)} \qquad s_i = \begin{pmatrix} 1 & 0 & 0 \\ 0 & \zeta + \alpha_i & \gamma_i \end{pmatrix} .$$

Then

$$\begin{aligned} s_0 T_{01} &= \begin{pmatrix} 1 & 0 & 0 & 0 & \cdots & 0 \\ 0 & z^{-1}(\zeta + \alpha_0) + \gamma_0^1 & z\gamma_0^1 & \gamma_0^2 & \cdots & \gamma_0^s \end{pmatrix} \\ &= \begin{pmatrix} 1 & 0 & 0 & 0 & \cdots & 0 \\ 0 & \zeta + \alpha_0 + z\gamma_0^1 & z^2\gamma_0^1 & z\gamma_0^2 & \cdots & z\gamma_0^s \end{pmatrix} \\ &= s_1 \ , \end{aligned}$$

and so

$$\text{(4.8)} \qquad \begin{cases} \alpha_1 = \alpha_0 + z\gamma_0^1 \\ \gamma_1^1 = z^2\gamma_0^1 \\ \gamma_1^t = z\gamma_0^t & t > 1 \ . \end{cases}$$

It follows from the transitions (4.6) and (4.8) that we can view $b = (b_0, b_1)$ as a section in $\Gamma\mathbb{F}_1$ and $\beta = (\beta_0, \beta_1)$ as a section in $\Gamma\mathbb{F}_{-1}$. Clearly, β corresponds to b under the reciprocal homeomorphism (2.4), which means that we can always think of both b and β as being defined (even if the coordinate funtions b_i or β_i were not already). For example, if β is the zero section (a holomorphically rigid condition according to §2.2), then b is the infinity section (also rigid). The projection $s \mapsto b$ is continuous, and so $\Gamma E_1(f)$ separates according to the separation of $\Gamma\mathbb{F}_1$ given in §2.2. We write Γ^∞ for the subspace of $\Gamma E_1(f)$ for which b is the infinity section, and we write Γ^r for the subspace for which $b \in \Gamma^r\mathbb{F}_1$. (We shall see in §4.4 that this separation is also precisely according to the second Chern class of the associated maps $\varphi : \mathbb{P}^1 \times \mathbb{P}^1 \to \mathbb{G}$.)

EXAMPLES 4.9. We take $\mathbb{G} = \mathbb{G}_{2,2}$; f is as in (0.4).

(a) For any polynomial $a_0 \in \mathbb{C}[w]$,

$$s_0 = \begin{pmatrix} \zeta + a_0 & 0 & w^2 - 1 & w(w^2-1) \\ w/(w^2-1) & 1 & 0 & 0 \end{pmatrix}$$

defines a rational map $U_0 \to \mathrm{Rat}_1(\mathbb{G})$. However, s_0 cannot be extended to give a section $s \in \Gamma E_1(f)$ because the transitions (4.6) cannot be satisfied. For example,

$$c_1^1(z) = zc_0^1(z^{-1}) = z^{-1} - z,$$

which is not holomorphic in z.

(b) Let

$$s_0 = \begin{pmatrix} \zeta + a_0 & 0 & w - 1 & 0 \\ w/(w-1) & 1 & 0 & 0 \end{pmatrix}.$$

The transitions (4.6) can be satisfied, but they place a constraint on a_0 and a_1:

$$a_1(z) = a_0(z^{-1}) - z \cdot \frac{z^{-1}}{z^{-1}-1} \cdot (z^{-1} - 1) = a_0(z^{-1}) - 1.$$

Since a_1 must be holomorphic in z and a_0 must be holomorphic in $w = z^{-1}$, it follows that $a_0 = a \in \mathbb{C}$ and $a_1 = a - 1$. Then

$$s_1 = \begin{pmatrix} \zeta + a - 1 & 0 & 1 - z & 0 \\ z/(1-z) & 1 & 0 & 0 \end{pmatrix}$$

extends s_0 to give a section $s = (s_0, s_1) \in \Gamma E_1(f)$. Since

$$b = \left(\frac{w}{w-1}, \frac{z}{1-z}\right) \in \Gamma^1\mathbb{F}_1,$$

it follows that $s \in \Gamma^1$.

We first analyze the subspace Γ^∞. In this case we have $\gamma = (\gamma_0, \gamma_1) \in \Gamma V_0$, where $V = \mathcal{O}(2) \oplus \mathcal{O}(1)^{s-1}$. Since α_0 and γ_0^1 are polynomials in z^{-1}, while α_1 is a polynomial in z, it follows from the transition (4.8) that α_0 and α_1 are both determined from γ^1 up to an additive complex number α. The section s determines uniquely and is determined by the associated data α and γ, and this correspondence gives a homeomorphism

$$\Gamma^\infty \cong \mathbb{C} \times \Gamma V_0. \tag{4.10}$$

In particular, Γ^∞ is a component of $\Gamma E_1(f)$.

For sections in Γ^r, the situation is only slightly more complicated. As in the case above, a_0 and a_1 are determined by b and c^1 up to an additive complex number a. Although $c = (c_0, c_1) \in \Gamma[\mathcal{O}(1) \oplus \mathcal{O}^{s-1}]$, the data b and c are redundant. We must account for the fact that the the section c vanishes whenever b has poles. Let B be a lift of b to $\Gamma[\mathcal{O}(r+1) \oplus \mathcal{O}(r)]_0$ and let $(p,q) \in \mathcal{P}(r+1, r)$ satisfy $B_1(z) = (p(z), q(z))$ as in §2.3. The lift B and the pair (p,q) are determined from b only up to multiples from $\mathbb{C}^*$. Since $b_1 = p/q$, we have that $q | c_1^t$ and $\gcd(c_1^t/q) = 1$. Notice that $r = \deg q$ must be ≤ 1. (If $r > 1$, then each component section c^t must have at least two zeros. But any holomorphic section of $\mathcal{O}(1)$ or $\mathcal{O}$ with at least two zeros must be identically zero, so it would follow that c is identically zero, a contradiction.) Furthermore, c_1/q determines a section

$$\xi \in \Gamma W_0,$$

where $W = \mathcal{O}(1-r) \oplus \mathcal{O}(-r)^{s-1}$. Another choice of lift gives an associated polynomial pair $(\lambda p, \lambda q) \in \mathcal{P}(r+1, r)$ and section $\lambda^{-1}\xi \in \Gamma W_0$ for some $\lambda \in \mathbb{C}^*$. Thus, the data b and c are equivalent to

$$\langle (p,q), \xi \rangle \in \mathcal{P}(r+1, r) \times_{\mathbb{C}^*} \Gamma W_0, \tag{4.11}$$

where the action of $\mathbb{C}^*$ identifies

$$\langle (\lambda p, \lambda q), \xi \rangle = \langle (p,q), \lambda \xi \rangle.$$

The correspondence of s with the data a and $\langle (p,q), \xi \rangle$ gives the homeomorphism

$$\Gamma^r \cong \mathbb{C} \times \left[\mathcal{P}(r+1, r) \times_{\mathbb{C}^*} \Gamma W_0 \right]. \tag{4.12}$$

4.3. $\Gamma E_1(f)$ for generic $f \in \mathrm{Rat}_k(\mathbb{G})$. The procedure of analysis is the same as in the previous subsection, except that we use the transition T_{01} from (3.24). We therefore assume that f has the form (3.18) with $s \geq k$ and that $n \geq m$. Suppose the component functions s_i have the form (4.5). Then

$$\begin{aligned} s_1 &= s_0 T_{01} \\ &= \begin{pmatrix} z^{-m}(\zeta + a_0) - N_{21}\omega_0 & N_{11}\omega_0 & zc_0^1 & \cdots & zc_0^k & c_0^{k+1} & \cdots & c_0^s \\ z^{-m} b_0 & z^{-n} & 0 & \cdots & 0 & 0 & \cdots & 0 \end{pmatrix} \\ &= \begin{pmatrix} \zeta + a_0 - (z^m N_{21} + z^n b_0 N_{11})\omega_0 & 0 & z^{m+1}c_0^1 & \cdots & z^m c_0^{k+1} & \cdots \\ z^{n-m} b_0 & 1 & 0 & \cdots & 0 & \cdots \end{pmatrix}, \end{aligned}$$

where we have written

$$\omega_0 = \sum_{t=1}^{k} z^{t-k} c_0^t, \tag{4.13}$$

and so

$$\begin{cases} a_1 = a_0 - (z^m N_{21} + z^n b_0 N_{11})\omega_0 & \\ b_1 = z^{n-m} b_0 & \\ c_1^t = z^{m+1} c_0^t & 1 \leq t \leq k \\ c_1^t = z^m c_0^t & k+1 \leq t \leq s\,. \end{cases} \tag{4.14}$$

If $\beta_i \equiv 0$, then as before s_i has the form (4.7). With

$$\vartheta_0 = \sum_{t=1}^{k} z^{t-k} \gamma_0^t \tag{4.15}$$

we obtain

$$
\begin{aligned}
s_1 &= s_0 T_{01} \\
&= \begin{pmatrix} z^{-m} & 0 & 0 & \cdots & 0 & 0 & \cdots & 0 \\ -N_{21}\vartheta_0 & z^{-n}(\zeta+\alpha_0)+N_{11}\vartheta_0 & z\gamma_0^1 & \cdots & z\gamma_0^k & \gamma_0^{k+1} & \cdots & \gamma_0^s \end{pmatrix} \\
&= \begin{pmatrix} 1 & 0 & 0 & \cdots & 0 & \cdots \\ 0 & \zeta+\alpha_0+z^n N_{11}\vartheta_0 & z^{n+1}\gamma_0^1 & \cdots & z^n\gamma_0^{k+1} & \cdots \end{pmatrix},
\end{aligned}
$$

and so

$$
(4.16) \qquad \begin{cases} \alpha_1 = \alpha_0 + z^n N_{11}\vartheta_0 & \\ \gamma_1^t = z^{n+1}\gamma_0^t & 1 \le t \le k \\ \gamma_1^t = z^n \gamma_0^t & k+1 \le t \le s\,. \end{cases}
$$

Let $\nu = n-m \geq 0$. As in the preceding subsection, we view $b = (b_0, b_1) \in \Gamma\mathbb{F}_\nu$, and $\beta = (\beta_0, \beta_1) \in \Gamma\mathbb{F}_{-\nu}$ corresponds under the homeomorphism (2.4). $\beta \equiv 0$ precisely when b is the infinity section, which is rigid when $\nu > 0$. In any case, $s \mapsto b$ is continuous and ensures separation of $\Gamma E_1(f)$ according to the separation in $\Gamma\mathbb{F}_\nu$. As before, we denote by Γ^r the subspace of sections $s \in \Gamma E_1(f)$ for which $b \in \Gamma^r\mathbb{F}_\nu$, and if $\nu > 0$ we denote by Γ^∞ the space of sections s for which b is the infinity section.

EXAMPLES 4.17. We take $\mathbb{G} = \mathbb{G}_{2,4}$ and we let f be as in §3.3.3 (viewing f as a map to $\mathbb{G}_{2,4}$ through the standard inclusion $\mathbb{G}_{2,3} \subset \mathbb{G}_{2,4}$). Recall that $m = 1$, $n = 2$, so that $\nu = 1$, and $N_{11} = -\frac{1}{2}z + 1$, $N_{21} = -z$.

(a) Let

$$
s_0 = \begin{pmatrix} \zeta + a_0 & 0 & w & w(w-1) & w(w+1) & w \\ (w-1)/w & 1 & 0 & 0 & 0 & 0 \end{pmatrix},
$$

where $a_0 \in \mathbb{C}[w]$. Then

$$
\omega_0 = z^{-2}w + z^{-1}w(w-1) + w(w+1) = 2w^3 + w,
$$

and so the transition (4.14) gives

$$
\begin{aligned}
a_1(z) &= a_0(z^{-1}) - \Big(z(-z) + z^2 \cdot \frac{w-1}{w} \cdot (-\tfrac{1}{2}z + 1)\Big)(2w^3 + w) \\
&= a_0(z^{-1}) - \tfrac{1}{2}z^3 + \tfrac{3}{2}z^2 - z + 3.
\end{aligned}
$$

Since a_1 is holomorphic in z and a_0 is holomorphic in $w = z^{-1}$, $a_0(w) = a \in \mathbb{C}$ and $a_1(z) = -\frac{1}{2}z^3 + \frac{3}{2}z^2 - z + 3 + a$. The polynomials c_0^t allow the remaining transitions in (4.14) to be satisfied, and s_0 can be extended with

$$
s_1 = \begin{pmatrix} \zeta + a_1 & 0 & z & 1-z & 1+z & 1 \\ z - z^2 & 1 & 0 & 0 & 0 & 0 \end{pmatrix}
$$

to give a section $s = (s_0, s_1) \in \Gamma E_1(f)$. Since

$$
b = \Big(\frac{w-1}{w}, z - z^2\Big) \in \Gamma^1\mathbb{F}_1,
$$

it follows that $s \in \Gamma^1$.

(b) Let

$$s_0 = \begin{pmatrix} \zeta + a_0 & 0 & w+2 & w(w-1) & w(w+1) & w+2 \\ w-1 & 1 & 0 & 0 & 0 & 0 \end{pmatrix}.$$

Then it is easy to check that

$$\omega_0 = 2w^3 + 2w^2 + w$$

and

$$a_1(z) = a_0(z^{-1}) - \left(2z^{-2} - 3z^{-1} - 3 - \tfrac{3}{2}z + \tfrac{1}{2}z^2\right).$$

Thus, to extend we must have $a_0(w) = 2w^2 - 3w + a$, where $a \in \mathbb{C}$, in which case $a_1(z) = -\frac{1}{2}z^2 + \frac{3}{2}z + 3 + a$. Then s_0 can be extended with

$$s_1 = \begin{pmatrix} \zeta + a_1 & 0 & z+2z^2 & 1-z & 1+z & 1+2z \\ 1-z & 1 & 0 & 0 & 0 & 0 \end{pmatrix}.$$

Since $b = (w-1, 1-z) \in \Gamma^0\mathbb{F}_1$, we have $s \in \Gamma^0$.

If $\nu > 0$, then following the analysis in the preceding subsection we obtain the homeomorphism

$$\Gamma^\infty \cong \mathbb{C} \times \Gamma V_0, \tag{4.18}$$

where $V = \mathcal{O}(n+1)^k \oplus \mathcal{O}(n)^{s-k}$. If $\nu > 0$ and $r > 0$ or if $\nu = 0$ and $r \geq 0$, then we obtain

$$\Gamma^r \cong \mathbb{C} \times \left[\mathcal{P}(r+\nu, r) \times_{\mathbb{C}^*} \Gamma W_0\right], \tag{4.19}$$

where $W = \mathcal{O}(m+1-r)^k \oplus \mathcal{O}(m-r)^{s-k}$. Notice that ΓW_0 is nonempty if and only if $r \leq m+1$. The spaces Γ^∞ and Γ^r are components of $\Gamma E_1(f)$.

4.4. The induced Chern class $\varphi^* c_2$. In this subsection we compute the Chern class $\varphi^* c_2$ induced by the a map $\varphi : \mathbb{P}^1 \times \mathbb{P}^1 \to \mathbb{G}$ associated to a section in $\Gamma E_1(f)$, where f is as in (3.18). (The map f in (0.4) is a special case; see the observation at the end of §3.3.2.) In particular, we shall see that the separation of $\Gamma E_1(f)$ into components is according to $\varphi^* c_2$. m and n are as in (3.14).

If $X \subset \mathbb{G}$ is a submanifold whose homology class is Poincaré dual to $c_2 \in H^4(\mathbb{G}; \mathbb{Z})$, then

$$\langle \varphi^* c_2, [\mathbb{P}^1 \times \mathbb{P}^1] \rangle = \langle c_2, [\varphi(\mathbb{P}^1 \times \mathbb{P}^1)] \rangle = [X] \cap [\varphi(\mathbb{P}^1 \times \mathbb{P}^1)], \tag{4.20}$$

where the last expression indicates the homology intersection pairing. c_2 is Poincaré dual to the submanifold

$$X = \left\{ \begin{pmatrix} 0 & * & \cdots & * \\ 0 & * & \cdots & * \end{pmatrix} \right\}$$

([**GrH**], p. 410). Thus, to compute $\varphi^* c_2$ we simply need to write the explicit map φ and count the total algebraic vanishing of the first column. Of course, any convenient section s within a component of $\Gamma E_1(f)$ can be used to compute the Chern class corresponding to that component.

Since $\varphi(z:w;1:0) = f(z:w)$, which does not vanish in the first column (see (3.18)), we need only consider vanishing at points $(z:w;\zeta:1) \in \mathbb{P}^1 \times \mathbb{C} \subset \mathbb{P}^1 \times \mathbb{P}^1$. It follows from (3.2) that

$$\begin{cases} \varphi(1:w;\zeta:1) = s_0(w)B_0(w) \\ \varphi(z:1;\zeta:1) = s_1(z)B_1(z) \,, \end{cases} \tag{4.21}$$

where B_0 and B_1 are as in (3.25) and (3.26). The variable ζ appears in the section values s_i as we have seen in §§4.2 and 4.3. Explicitly, we have

$$\varphi(1:w;\zeta:1) = \begin{cases} \begin{pmatrix} (\zeta+a_0)M_{11} & \cdots \\ M_{11}b_0 + M_{21} & \cdots \end{pmatrix} & \text{(4.22 a)} \\ \begin{pmatrix} M_{11}+\beta_0 M_{21} & \cdots \\ (\zeta+\alpha_0)M_{21} & \cdots \end{pmatrix} & \text{(4.22 b)} \end{cases}$$

and

$$\varphi(z:1;\zeta:1) = \begin{cases} \begin{pmatrix} (\zeta+a_1)N_{11} & \cdots \\ N_{11}b_1 + N_{21} & \cdots \end{pmatrix} & \text{(4.23 a)} \\ \begin{pmatrix} N_{11}+\beta_1 N_{21} & \cdots \\ (\zeta+\alpha_1)N_{21} & \cdots \end{pmatrix} & \text{(4.23 b)} \end{cases}$$

Recall from (3.22) that $(N_{11}, N_{21}) \in \mathcal{P}(m,n)$, so $\gcd(N_{11}, N_{21}) = 1$ and either $\deg N_{11} = m$ or $\deg N_{21} = n$. Also, $M_{11} = w^m N_{11}$, $M_{21} = w^n M_{21}$, and $\gcd(M_{11}, M_{21}) = 1$.

PROPOSITION 4.24. *If b is the infinity section, then $\varphi^* c_2 = m$. If $b \in \Gamma^r \mathbb{F}_k$, then $\varphi^* c_2 = r + n$.*

PROOF. We need to count the total algebraic vanishing of φ in the first column. First assume that b is the infinity section. Then $\beta \equiv 0$, so we must use the forms (4.22 b) and (4.23 b). For z finite we count the the intersection index

$$\#[N_{11} \cap (\zeta+\alpha_1)N_{21}] = \#[N_{11} \cap (\zeta+\alpha_1)],$$

and since N_{11} and $\zeta+\alpha_1$ clearly have distinct tangents, we accumulate $\deg N_{11}$ toward the total vanishing. At $w=0$, we count

$$\#[M_{11} \cap (\zeta+\alpha_0)M_{21}] = \#[N_{11} \cap (\zeta+\alpha_1)],$$

which is the order of vanishing of M_{11} at $w=0$. Since $M_{11} = w^m N_{11}$, this order of vanishing and $\deg N_{11}$ add to give m.

Now assume that $b \in \Gamma^r \mathbb{F}_k$. We consider first the case $\deg N_{11} = m$. Since $M_{11} = w^m N_{11}$, it follows that $M_{11}(0) \neq 0$. We write $b_1 = p/q$, where we are free to put $q=1$. Thus, $\deg p = r + \nu$ and β vanishes to order r at $w=0$. For z finite, we use (4.23 a) to count the vanishing from the intersection index

$$\#[(\zeta+a_1)N_{11} \cap (N_{11}b_1 + N_{21})] = \#[(\zeta+a_1) \cap (N_{11}p + N_{21})] = r+n,$$

which is the degree of the polynomial $N_{11}p + N_{21}$. It remains to count any contribution at $w = 0$. If $r > 0$, then in (4.22 b) $M_{11}(0) + \beta_0(0)M_{21}(0) = M_{11}(0) \neq 0$, so there is no contribution at $w = 0$. If $r = 0$, then we are free to choose β_0 to ensure that again we get no contribution at $w = 0$. Thus, if $\deg N_{11} = m$, then the total vanishing of the first column of φ is $r + n$. Finally, if $\deg N_{21} = n$, then we argue similarly, assuming $\beta_1 = q/p$ with $p = 1$ and using (4.22 a) and (4.23 b), to count the intersections. □

5. Basing and the theorems

In this section we complete the work of proving Theorems A and B by analyzing the effect of imposing the basing condition (3.4) of the map g on the section spaces Γ^∞, Γ^r of §4.

5.1. Basing for $f \in \mathrm{Rat}_1(\mathbb{G})$ and Theorem B. For Theorem B we assume f is of degree one and $s > 1$. By post-composing with an automorphism of $\mathbb{G}$, we may assume further that f is as in (0.4). We have seen that the components

$$\Gamma^\infty,\ \Gamma^0,\ \Gamma^1$$

of $\Gamma E_1(f)$ correspond to the Chern classes $c_2 = 0, 1, 2$ (resp.). It remains to analyze the structure of the associated based spaces $\Gamma^\infty_* \cong \rho_0^{-1}(f, g)$ and $\Gamma^r_* \cong \rho_{r+1}^{-1}(f, g)$, where the basing is imposed from the map g according to (3.4). Recall that the basing is imposed at $w = 0$, and specifically

$$g = s_0(0)B_0(0).$$

From (3.15), we see that $B_0(0)$ is the identity, and so $g = s_0(0)$.

Let us consider first sections $s \in \Gamma^\infty_*$. For such a section, $\beta \equiv 0$, and so it is necessary that $g \in \mathcal{U}_{0,1} - \mathcal{U}_{1,0}$. Recall that s is determined by the data $\alpha \times \gamma$ in

$$\mathbb{C} \times \Gamma V_0,$$

where $V = \mathcal{O}(2) \oplus \mathcal{O}(1)^{s-1}$. The value of α_0 at $w = 0$ is determined by g, and once γ is specified, the complex number α is determined. The value $\gamma_0(0)$ is determined by g, and so the freedom in selecting $s \in \Gamma^\infty_*$ is the same as the freedom in selecting the based section $\gamma \in \Gamma_* V_0$. If $g \notin \mathrm{Rat}_1(\mathbb{G}_{2,1})$ (i.e., if the image of g does not lie in $\mathbb{G}_{2,1} \subset \mathbb{G}$), then $\gamma_0^t(0) \neq 0$ for some $t > 1$. It follows that γ_1^t is of degree one, and so after reordering coordinates, the basing for γ satisfies the hypotheses of Lemma 2.16. The based section γ is therefore equivalent to an element in

$$\mathrm{Rat}_1(\mathbb{P}^{s-1}) \times \mathbb{C}. \tag{5.1}$$

On the other hand, if $g \in \mathrm{Rat}_1(\mathbb{G}_{2,1})$, then $\gamma_0^t(0) = 0$ for $t > 1$, while $\gamma_0^1(0) \neq 0$ is determined by g. It follows that $(\gamma_1^2, \ldots, \gamma_1^s) \in \mathbb{C}^{s-1} - \{0\}$, while γ_1^1 is a degree two polynomial whose leading coefficient is determined and whose other two coefficients are free. Thus, the based section γ is equivalent to an element of

$$(5.2) \qquad (\mathbb{C}^{s-1} - \{0\}) \times \mathbb{C}^2.$$

Since (5.1) and (5.2) are biholomorphic, we have the following

PROPOSITION 5.3. *Let f be as in (0.4). If $g \in \mathcal{U}_{0,1} - \mathcal{U}_{1,0}$, then*

$$\rho_0^{-1}(f,g) \cong \mathrm{Rat}_1(\mathbb{P}^{s-1}) \times \mathbb{C},$$

and if $g \in \mathcal{U}_{1,0}$, then $\rho_0^{-1}(f,g)$ is empty.

Next, assume that $s \in \Gamma^1_*$. Then $b \in \Gamma^0\mathbb{F}_1$, so b has no poles and it is necessary that $g \in \mathcal{U}_{1,0}$. From (4.12), s is determined by $a \times \langle (p,q), \xi \rangle$ in

$$\mathbb{C} \times \left[\mathcal{P}(1,0) \times_{\mathbb{C}^*} \Gamma W_0\right],$$

where $W = \mathcal{O}(1) \times \mathcal{O}^{s-1}$. As above, g together with $\langle (p,q), \xi \rangle$ determine a. q is a nonzero constant, so it can be normalized to 1. Then $b_0(0) = wp(w^{-1})|_{w=0}$ determines the coefficient of z in p, while the other coefficient in p is free. Since $q = 1$, we obtain a preferred choice of $\xi = c_1$. Since all components of c_1 but the first are constant, it follows that at least one of the constants must be nonzero. This implies that $g \notin \mathrm{Rat}_1(\mathbb{G}_{2,1})$ and that g determines c_1^t for $t > 1$. The coefficient of z in c_1^1 is determined by g, while the other coefficient is free. In total there are two free coefficients, and so we have the

PROPOSITION 5.4. *Let f be as in (0.4). If $g \in \mathcal{U}_{1,0}$ and $g \notin \mathrm{Rat}_1(\mathbb{G}_{2,1})$, then*

$$\rho_1^{-1}(f,g) \cong \mathbb{C}^2.$$

Otherwise, $\rho_1^{-1}(f,g)$ is empty.

Finally we consider $s \in \Gamma^1_*$. From (4.12), s is given by data $a \times \langle (p,q), \xi \rangle$. As above a is determined from g and the other data, and

$$\langle (p,q), \xi \rangle \in \mathcal{P}(2,1) \times_{\mathbb{C}^*} \Gamma W_0,$$

where $W = \mathcal{O} \oplus \mathcal{O}(-1)^{s-1}$. A holomorphic section of W must vanish in the $\mathcal{O}(-1)$ summands, so we can view $\xi \in \Gamma\mathcal{O}_0$. Thus, it is necessary that $g \in \mathrm{Rat}_1(\mathbb{G}_{2,1})$. Given a polynomial pair (p,q) corresponding to a lift of b_1, the map g then determines $\xi = c_1/q = -\gamma_1/p$, so the only freedom in determining the section s is in choosing $b \in \Gamma^1\mathbb{F}_1$. b is based at $w = 0$ from g, and as we saw in Lemma 2.13, $\Gamma^1_*\mathbb{F}_1 \cong \mathbb{C}^2 \times \mathbb{C}^*$. This proves the

PROPOSITION 5.5. *Let f be as in (0.4). If $g \in \mathrm{Rat}_1(\mathbb{G}_{2,1})$, then*

$$\rho_2^{-1}(f,g) \cong \mathbb{C}^2 \times \mathbb{C}^*.$$

Otherwise, $\rho_2^{-1}(f,g)$ is empty.

Combining Propositions 5.3, 5.4, and 5.5 gives Theorem B. Notice that only the conditions $g \in \mathcal{U}_{1,0}$ and $g \notin \mathrm{Rat}_1(\mathbb{G}_{2,1})$ of Proposition 5.4 apply to generic g.

5.2. Basing for $f \in \mathrm{Rat}_k(\mathbb{G})$ and Theorem A. For Theorem A, f and g can be taken to satisfy suitable generic conditions. We assume that f is generic as in §3.3.2, and so by using an automorphism of $\mathbb{G}$ we can assume f has the form (3.18) with $s > k$. We take m, n, and ν as in §4.3. According to (3.4), $g = s_0(0)B_0(0)$ and from (3.25)

$$B_0(0) = \begin{pmatrix} M & 0 \\ 0 & id_s \end{pmatrix},$$

where M is determined from f. Right multiplication by $B_0(0)$ gives an automorphism of $\mathrm{Rat}_1(\mathbb{G})$, so generic conditions on g correspond to generic conditions on $\bar{g} = s_0(0)$. In particular, we assume that

$$\bar{g} \in \mathcal{U}_{1,0} \tag{5.6}$$

and

$$\bar{g} \notin \mathrm{Rat}_1(\mathbb{G}_{2,k}). \tag{5.7}$$

Condition (5.6) implies that b is not the infinity section, so we do not obtain sections s in Γ^∞. Thus, $s \in \Gamma_*^r$ is determined according to (4.19) by the data

$$a \times \langle (p,q), \xi \rangle \in \mathbb{C} \times \left[\mathcal{P}(\nu + r, r) \times_{\mathbb{C}^*} \Gamma W_0 \right],$$

where $W = \mathcal{O}(m+1-r)^k \oplus \mathcal{O}(m-r)^{s-k}$. As we saw in §5.1, a is determined by $\bar{g}$ and the other data. The condition (5.7) forces $r \le m$, since otherwise the section ξ must vanish in the last $s-k$ coordinates. In terms of the Chern class $c = c_2 = r + n$, we see that $n \le c \le k$. Furthermore, (5.6) implies that $\deg q = r$ (i.e., b does not have a pole at $w = 0$). Thus, q can be normalized to be monic, giving preferred choices of $p = b_1 q$ and $\xi = c_1/q$. According to (2.11), the pair (p,q) can be chosen freely in

$$\mathrm{Rat}_r \times \mathbb{C}^\nu. \tag{5.8}$$

The section

$$\xi \in \Gamma_* \left[\mathcal{O}(m+1-r)^k \oplus \mathcal{O}(m-r)^{s--k} \right]_0$$

is based from $\bar{g}$, and (5.7) ensures that (after reordering) the basing condition of Lemma 2.16 is satisfied. Thus, ξ can be chosen freely in

$$\mathrm{Rat}_{m-r}(\mathbb{P}^{s-1}) \times \mathbb{C}^k. \tag{5.9}$$

Combining (5.8) and (5.9) and using $r = c - n$, $m - r = k - c$, $\nu + k = 2n$, we obtain Theorem A. In particular, for f as in (3.18), suitable generic conditions on g are obtained from the conditions (5.6) and (5.7) on $\bar{g}$.

REFERENCES

[A] M. F. Atiyah, *Instantons in two and four dimensions*, Comm. Math. Phys. **93** (1984), 437–451.

[ByH] C. I. Byrnes and N. E. Hurt, *On the moduli of linear dynamical systems*, Studies in Analysis, Advances in Mathematics Supplementary Studies, v. 4, 1979, pp. 83–122.

[C] J. M. C. Clark, *The consistent selection of parameterizations in system identification*, Proc. of the Joint Automatic Control Conference, 1976, pp. 576–580.

[CCMM] F. R. Cohen, R. L. Cohen, B. M. Mann, and R. J. Milgram, *The topology of rational functions and divisors of surfaces*, Acta Math. **166** (1991), 163–221.

[CS] R. L. Cohen and D. H. Shimamoto, *Rational functions, labelled configurations, and Hilbert schemes*, J. London Math. Soc. **43** (1991), 509–528.

[D] S. K. Donaldson, *Instantons and geometric invariant theory*, Comm. Math. Phys. **93** (1984), 453–461.

[F] W. Fulton, *Algebraic Curves*, Addison-Wesley, Redwood City, Calif., 1989.

[Gr] J. Gravesen, *On the topology of spaces of holomorphic maps*, Acta Math. **162** (1989), 247–286.

[GrA] P. Griffiths and J. Adams, *Topics in Algebraic and Analytic Geometry*, Mathematical Notes no. 13, Princeton University Press, Princeton, 1974.

[GrH] P. Griffiths and J. Harris, *Principles of Algebraic Geometry*, John Wiley & Sons, New York, 1978.

[Gu] M. A. Guest, *Topology of the space of absolute minima of the energy functional*, Amer. J. of Math **106** (1984), 21–42.

[H1] J. W. Havlicek, *On spaces of holomorphic maps from two copies of the Riemann sphere to complex Grassmannians*, Ph.D. thesis, Stanford University, 1992.

[H2] J. W. Havlicek, *The cohomology of holomorphic self maps of the Riemann sphere*, Math. Zeit. (to appear).

[Hu] J. Hurtubise, *Instantons and jumping lines*, Comm. Math. Phys. **105** (1986), 107–122.

[K] F. C. Kirwan, *On spaces of maps from Riemann surfaces to Grassmannians and applications to the cohomology of moduli of vector bundles*, Ark. Mat. **24** (1986), 221–275.

[MM1] B. M. Mann and R. J. Milgram, *Some spaces of holomorphic maps to complex Grassmann manifolds*, Jour. Diff. Geom. **33** (1991), 301–324.

[MM2] ———, *The topology of rational maps to Grassmannians and a homotopy theoretic proof of the Kirwan stability theorem*, Preprint, University of New Mexico and Stanford University (1991).

[MM3] ———, *On the moduli space of $SU(n)$ monopoles and holomorphic maps to flag manifolds*, Preprint, University of New Mexico and Stanford University (1991).

[S] G. Segal, *The topology of spaces of rational functions*, Acta Math. **143** (1979), 39–72.

[Ser] J.-P. Serre, *Géometrie algébrique et géométrie analytique*, Ann. Inst. Fourier **6** (1956), 1–42.

[T] B. Totaro, *The cohomology ring of the space of rational functions*, Preprint, MSRI (1990).

DEPARTMENT OF MATHEMATICS, MICHIGAN STATE UNIVERSITY, EAST LANSING, MICHIGAN, 48824-1027

E-mail address: havlicek@math.msu.edu

Contemporary Mathematics
Volume **182**, 1995

A Real Algebraic Vector Bundle is Strongly Algebraic whenever its Total Space is Affine

J. HUISMAN

ABSTRACT. The aim of this note is to prove the following statement. Let X be an affine real algebraic variety and let $\xi = (E, \pi)$ be a real algebraic vector bundle over X. Then, ξ is strongly algebraic if and only if E is an affine real algebraic variety.

The aim of this note is to prove the following statement in a more elementary way than that of [**3**].

THEOREM. *Let X be an affine real algebraic variety*[1] *and let $\xi = (E, \pi)$ be a real algebraic vector bundle over X. Then, ξ is strongly algebraic if and only if E is an affine real algebraic variety.*

The only-if part being trivial, the difficulty lies in showing that ξ is strongly algebraic if E is affine. Some effort has already been put into this problem, see [**5**], [**3**]. Both papers follow the idea of enlarging the scheme $\operatorname{Spec}\mathcal{R}(E)$ as a scheme over $\operatorname{Spec}\mathcal{R}(X)$ in order to make it a geometric vector bundle over $\operatorname{Spec}\mathcal{R}(X)$. While the former paper uses arguments which I could not follow, the latter applies a generalisation, due to M. Artin, of a theorem of A. Weil on rational group laws [**1**].

We will however prove the Theorem in a more elementary way. The idea roughly is the following. The normal bundle of the zero section in E is strongly algebraic[2], E being affine. This normal bundle should be isomorphic, at least morally, to E as a real algebraic vector bundle.

1991 *Mathematics Subject Classification.* 14P05.

The author is supported by the Netherlands Organisation for Scientific Research (NWO).

The author would like to thank Institut Fourier, and C. Peters in particular, for the warm hospitality he recieved during his stay.

[1]In this note definitions and notations from [**2**] are freely used.

[2]I am grateful to S. Akbulut for stimulating me to study more closely this normal bundle.

The actual proof turns out to be more agreeable if we work with the conormal bundle instead of the normal one, and with the sheaf of sections instead of the vector bundles. The idea of the proof remains the same, however.

Recall that a real algebraic vector bundle $\xi = (E, \pi)$ over an affine real algebraic variety X is strongly algebraic if and only if there exists a projective $\mathcal{R}(X)$-module M, of finite type, such that the sheaf $\widetilde{M} = M \otimes_{\mathcal{R}(X)} \mathcal{R}_X$ is isomorphic to the sheaf of sections of ξ ([**2**], Théorème 12.1.7). Actually, the only obstruction for ξ to be strongly algebraic is the nonexistence of an $\mathcal{R}(X)$-module M such that $\widetilde{M}$ is isomorphic to the sheaf of sections of ξ. This is formulated in the next Lemma. Although its proof is standard we include it for the reader's convenience.

LEMMA. *Let $\xi = (E, \pi)$ be a real algebraic vector bundle on the affine real algebraic variety X and let M be an $\mathcal{R}(X)$-module. If $\widetilde{M}$ is isomorphic to the sheaf of sections of ξ then M is projective and of finite type. In particular, ξ is strongly algebraic.*

PROOF. There exists a finite open covering $\{U_i\}$ of X such that $\widetilde{M}_{|U_i}$ is a free sheaf of $\mathcal{R}_{U_i}$-modules of finite rank. Choose finitely many

$$s_{ij} \in \Gamma(U_i, \widetilde{M}_{|U_i})$$

such that $\{s_{ij}\}_j$ generates $\Gamma(U_i, \widetilde{M}_{|U_i})$, for all i. We may assume that all sections s_{ij} extend to global sections of $\widetilde{M}$, again denoted by s_{ij}. We claim that $\{s_{ij}\}_{ij}$ generates M. Indeed, if $s \in M$ then there exist $f_i \in \mathcal{R}(X)$, with $f_i(x) \neq 0$ for all $x \in U_i$, and there exist $f_{ij} \in \mathcal{R}(X)$ such that

$$f_i s = \sum_j f_{ij} s_{ij},$$

for all i. Since $\bigcap f_i^{-1}(0) = \emptyset$, there exist $\lambda_i \in \mathcal{R}(X)$ such that $\sum \lambda_i f_i = 1$. Hence

$$s = \sum_i \lambda_i f_i s = \sum_{i,j} \lambda_i f_{ij} s_{ij}.$$

Thus $\{s_{ij}\}_{ij}$ generates M and, therefore, M is of finite type. Since M also is locally free, it follows that M is projective ([**4**], Corollary IV.3.6).

DEFINITION. Let X be a closed real algebraic subvariety of the real algebraic variety Y. Then the conormal sheaf $\mathcal{C}_{X/Y}$ of X in Y is the sheaf of $\mathcal{R}_X$-modules $\mathcal{I}/\mathcal{I}^2$, where $\mathcal{I}$ is the sheaf on Y of vanishing ideals of X, i.e.

$$\Gamma(U, \mathcal{I}) = \{f \in \Gamma(U, \mathcal{R}_Y) \mid \forall x \in U \cap X : f(x) = 0\},$$

where U is an open subset of Y.

Note that, a priori, $\mathcal{I}/\mathcal{I}^2$ only is a sheaf on Y. However, its support is obviously equal to X. Therefore, it might be considered to be a sheaf on X, and as such naturally has the structure of a sheaf of $\mathcal{R}_X$-modules.

PROOF OF THE THEOREM. Suppose E is affine. Embed X in E via the zero section. Then, the conormal sheaf $\mathcal{C}_{X/E}$ of X in E is isomorphic to the dual of the sheaf of sections $\Gamma(\cdot,\xi)$ of ξ. Indeed, let $\xi^\vee$ be the dual vector bundle of ξ. One constructs a morphism of sheaves of $\mathcal{R}_X$-modules

$$\varphi\colon \Gamma(\cdot,\xi^\vee) \longrightarrow \mathcal{C}_{X/E}$$

as follows. Let U be an open subset of X and s a section of $\xi^\vee$ over U. Define a regular function f_s on $\pi^{-1}(U)$ by $f_s(x,v) = s(x)v$. Since f_s vanishes on U (remember that we have embedded X in E via the zero section), f_s determines a section $\varphi(s)$ of $\mathcal{C}_{X/E}$ over U. This defines the morphism φ. Local triviality of ξ makes it easy to see that φ is in fact an isomorphism. Therefore, $\mathcal{C}_{X/E}$ is isomorphic to the sheaf of sections of the real algebraic vector bundle $\xi^\vee$.

Now, we are going to prove that $\xi^\vee$ is strongly algebraic, which obviously suffices in order to conclude the Theorem. Since E is affine, the sheaf of vanishing ideals $\mathcal{I}$ of X in E is isomorphic to $I \otimes_{\mathcal{R}(E)} \mathcal{R}_E$, where $I = \Gamma(E,\mathcal{I})$. Hence

$$\mathcal{C}_{X/E} \cong (I/I^2) \otimes_{\mathcal{R}(X)} \mathcal{R}_X.$$

It follows from the Lemma that $\xi^\vee$ is strongly algebraic. This finishes the proof.

REFERENCES

1. M. Artin, *Théorème de Weil sur la construction d'un groupe à partir d'une loi rationelle*, Schémas en Groupes II (SGA 3) (M. Demazure, A. Grothendieck, eds.), Lect. Notes Math., vol. 152, Springer Verlag, Berlin Heidelberg New York, 1970.
2. J. Bochnak, M. Coste, M.-F. Roy, *Géométrie algébrique réelle*, Ergeb. Math. Grenzgeb., 3. Folge, vol. 12, Springer Verlag, Berlin Heidelberg New York, 1987.
3. J. Huisman, *On real algebraic vector bundles* (to appear in Math. Z.).
4. E. Kunz, *Introduction to commutative algebra and algebraic geometry*, Birkhäuser, Boston Basel Stuttgart, 1985.
5. M. G. Marinari, M. Raimondo, *Fibrati vettoriali su varietà algebriche definite su corpi non algebricamente chiusi*, Boll. Unione Mat. Ital., V Ser., A **16** (1979), 128–136.

MATHEMATICAL INSTITUTE, UNIVERSITY OF UTRECHT, POSTBUS 80010, NL-3508 TA UTRECHT, THE NETHERLANDS

Current address: Institut Mathématique, Université de Rennes I, Campus de Beaulieu, 35042 Rennes Cedex, France

E-mail address: huisman@univ-rennes1.fr

Contemporary Mathematics
Volume **182**, 1995

Real Algebraic Principal Abelian Fibrations

Yildiray Ozan

Abstract

Let X be a nonsingular compact real algebraic set and $G = (\mathbb{Z}_2)^k \times H$ where H is an odd order Abelian group. In this paper, we will study algebraic principal G fibrations over the real algebraic set X and define the algebraic first cohomology group of X, $H^1_A(X, G)$, and show that the quotient group $H^1(X, G)/H^1_A(X, G)$ carries the obstruction to realize smooth principal G fibrations over X algebraically.

1. Introduction

The problem of real algebraic realization of topological or smooth objects has been studied by many authors. In [20], Seifert showed that any closed smooth submanifold $M \subseteq \mathbb{R}^n$ with trivial normal bundle is isotopic to a nonsingular component of a real algebraic set $X \subseteq \mathbb{R}^n$. Nash showed that every closed smooth manifold is diffeomorphic to a component of a nonsingular real algebraic set in some $\mathbb{R}^N$ ([16]). Later, Tognoli proved that any closed smooth manifold is diffeomorphic to a nonsingular real algebraic set in some $\mathbb{R}^N$ ([21]). In [2, 3] Akbulut and King improved Tognoli's result by showing that any closed smooth submanifold of $\mathbb{R}^n$ can be isotoped to the nonsingular points of an algebraic subset of $\mathbb{R}^n$. Algebraic realization of smooth manifolds and vector bundles with group actions have been studied by Dovermann and Masuda ([10]). In this paper, we will focus on the problem of *algebraic realization of smooth principal G fibrations over real algebraic sets.* For the fibrations we will study in this note G will denote $(\mathbb{Z}_2)^k \times H$ where H is an odd order Abelian group because our methods do not provide a universal algebraic principal fibration for Abelian 2-groups other than $\mathbb{Z}_2$ (see Remark 1 at the end of Section 3).

Let $X \subseteq \mathbb{R}^n$ and $Z \subseteq \mathbb{R}^m$ be semialgebraic sets. A map $F : X \longrightarrow Z$ is said to be entire rational if there exists $f_i, g_i \in \mathbf{R}[x_1, \ldots, x_n]$, $i = 1, \ldots, m$, such that each g_i vanishes nowhere on X and $F = (f_1/g_1, \ldots, f_m/g_m)$.

1991 *Mathematics Subject Classification.* Primary 14P25, 14E20; Secondary 57M10.

We say that X and Z are isomorphic to each other if there are entire rational maps $F : X \longrightarrow Z$ and $G : Z \longrightarrow X$ such that $F \circ G = id_Z$ and $G \circ F = id_X$.

2. Algebraic quotients

Let $X \subseteq \mathbb{R}^n$ be a semialgebraic set and G be a finite group acting on X. Then G is said to be acting algebraically on X if for each $g \in G$ the map $g : X \longrightarrow X$, $x \longmapsto g \cdot x$ is the restriction of some polynomial map $P_g : \mathbb{R}^n \longrightarrow \mathbb{R}^n$.

We will refer the reader to [1, 6] for the definitions of the concepts like irreducible, singular and nonsingular algebraic sets, ring of polynomial functions on X, complexification of X, etc.

Let $X \subseteq \mathbb{R}^n$ be a real algebraic set and G be a finite group acting algebraically on X. Since the action is given by polynomial maps $P_g : \mathbb{R}^n \longrightarrow \mathbb{R}^n$ the action extends to the complexification $X_{\mathbb{C}} \subseteq \mathbb{C}^n$ of X and it is given by the same polynomial maps. Let $J(X) \subseteq \mathbb{R}[x_1, \ldots, x_n]$ be the defining ideal of X and $S = \mathbb{R}[x_1, \ldots, x_n]/J(X)$ the ring of polynomial functions on X. We define a G action on S as follows: For $g \in G$ and $f \in S$, let $g \cdot f = f \circ g^{-1}$. Consider the subring T of S defined by

$$T = S^G = \{f \in S : g \cdot f = f, \forall g \in G\}.$$

Both T and S are $\mathbb{R}$ algebras. Moreover, it is well known that S is an algebraic extension of T and therefore, T is also a finitely generated $\mathbb{R}$ algebra. Say T is generated by $y_1, \ldots, y_m$ over the reals. Consider the complexifications $T_{\mathbb{C}}$ and $S_{\mathbb{C}}$ of T and S defined by

$$T_{\mathbb{C}} = T \otimes_{\mathbb{R}} \mathbb{C} \quad \text{and} \quad S_{\mathbb{C}} = S \otimes_{\mathbb{R}} \mathbb{C}.$$

It follows that these are finitely generated $\mathbb{C}$ algebras and $S_{\mathbb{C}}$ is nothing but the ring of complex polynomial functions on the complexification $X_{\mathbb{C}}$ of X ([17, 18, 19]).

The above action of G on S extends to $S_{\mathbb{C}}$, linearly over $\mathbb{C}$. With this definition of G action, immediately, we have that $S^G_{\mathbb{C}} = T_{\mathbb{C}}$.

Consider the maps $F_{\mathbb{C}} : X_{\mathbb{C}} \longrightarrow \mathbb{C}^m$ and $F = F_{\mathbb{C}|X} : X \longrightarrow \mathbb{R}^m$, where both are given by

$$x \longmapsto (y_1(x), \ldots, y_m(x)).$$

Note that these maps send G orbits to points because the y_i's are G equivariant polynomials.

Let Z denote the complex algebraic set corresponding to the $\mathbb{C}$ algebra $T_{\mathbb{C}}$ (i.e. the embedding of the maximal spectrum of $T_{\mathbb{C}}$ into $\mathbb{C}^m$ via $F_{\mathbb{C}}$). Then $F_{\mathbb{C}} : X_{\mathbb{C}} \longrightarrow Z$ is the morphism of the complex algebraic sets corresponding to the inclusion of $\mathbb{C}$ algebras $i : T_{\mathbb{C}} \longrightarrow S_{\mathbb{C}}$.

Let $Y_0 = F(X)$, which is a semialgebraic set, and let $Y \subseteq \mathbb{R}^m$ be the Zariski closure of Y_0. First of all, we have that $Z = Y_{\mathbb{C}}$ (Lemma 1.5 in [17]). Furthermore, the maps $F_{\mathbb{C}} : X_{\mathbb{C}} \longrightarrow Z = Y_{\mathbb{C}}$ and $F : X \longrightarrow Y_0$ are open, closed and proper maps. In particular they induce homeomorphisms between the quotient spaces $X_{\mathbb{C}}/G$ and X/G and the image sets $Y_{\mathbb{C}}$ and Y_0 respectively. Moreover, if the G action is free and X is nonsingular then the induced map $X/G \longrightarrow Y_0$ becomes a diffeomorphism.

In the literature, Y is usually denoted by $X//G$, the algebraic quotient of X by G. Schwarz's description of $X//G$ is as follows: Let I be the ideal of relations of the y_i's in $\mathbb{R}[z_1, \ldots, z_m]$. Then $X//G$ is the corresponding algebraic subset of $\mathbb{R}^m$ ([19]).

Now we will state some theorems which compare the semialgebraic set Y_0, which is the topological quotient X/G, with the algebraic quotient $Y = X//G$. See [17, 18, 19] for the proofs of these theorems and other statements made in this section.

THEOREM 2.1. *Let X be a real algebraic set and G is a finite group acting algebraically on X. Let Y be the Zariski closure of Y_0. Then:*

a) if $|G|$ is odd, then $Y_0 = Y$; i.e. Y_0 is algebraic;

b) if G acts freely on X and X is nonsingular, then Y_0 is a union of topological components of $Nonsing(Y)$ (G might have even order);

c) if $|G|$ is odd, X is nonsingular and G acts freely on X, then Y_0 is a nonsingular algebraic set.

In general, parts (a) and (c) of the above theorem do not hold for even order groups. For the $\mathbb{Z}_2$ case we have the following theorem.

For a proof of the following theorem see the proof of Theorem 1.9 in [17].

THEOREM 2.2. *Let $\mathbb{Z}_2 =< g >$ act algebraically and freely on a k dimensional nonsingular real algebraic set $X \subseteq \mathbb{R}^n$. Let W be the set*

$$\{p \in X_{\mathbb{C}} : g(p) = \bar{p}\}$$

where $\bar{\ }$ denotes the complex conjugation. Then, W is a real algebraic set of dimension less than or equal to k and the Zariski closure Y of Y_0 is the disjoint union of Y_0 and $F_{\mathbb{C}}(W)$.

The following remark will be used in Section 3 to construct a universal algebraic $\mathbb{Z}_2$ fibration.

REMARK. Assume that the above $\mathbb{Z}_2 =< g >$ action is linear. So by a linear change of coordinates we can assume that

$$g(x_1, \ldots, x_n) = (x_1, \ldots, x_{j-1}, -x_j, \ldots, -x_n)$$

for some $1 \leq j \leq n$. In this case W is equal to $X_{\mathbb{C}} \cap V_g$ where V_g is the linear subspace of $\mathbb{R}^{2n} = \mathbb{C}^n$ given by

$$x_l = 0, \quad l = j, \ldots, n \quad \text{and} \quad y_l = 0, \quad l = 1, \ldots, j-1,$$

where $(x_1, \ldots, x_n, y_1, \ldots, y_n) \in \mathbb{R}^{2n}(= \mathbb{C}^n$ by $z_l = x_l + iy_l)$. By identifying

$$(x_1, \ldots, x_{j-1}, 0, \ldots, 0, y_j, \ldots, y_n) \text{ with } (x_1, \ldots, x_{j-1}, y_j, \ldots, y_n),$$

we can identify W with the algebraic subset $W' \subseteq \mathbb{R}^n$, where

$$W' = \{(X, Y) = (x_1, \ldots, x_{j-1}, y_j, \ldots, y_n) : f(X, iY) = 0, \quad \forall f \in J(X^k)\}.$$

As an example, let $X = S^k \subseteq \mathbb{R}^{k+1}$ be the standard k-sphere and g the antipodal map. Then, $W' = \{(y_1, \ldots, y_{k+1}) \in \mathbb{R}^{k+1} : (iy_1)^2 + \cdots + (iy_{k+1})^2 = 1\} = \emptyset$. (The coordinate functions of F are $x_i x_j$ for $i \leq j = 1, \ldots, k+1$.) So $\mathbb{R}P^k$ is not only the geometric but also the algebraic quotient of S^k by the antipodal map. Indeed, the geometric and the algebraic quotients coincide also for the product spaces $Y \times S^k$ where Y is any algebraic set and $\mathbb{Z}_2$ fixes the first coordinate but acts on the second coordinate as the antipodal map.

3. An algebraic model for $K(G,1)$ and the first algebraic cohomology of X

3.1. Algebraic vector bundles. From now on the phrase *G fibration* will actually mean *principal G fibration*. Let X be a finite CW complex and G be a finite Abelian group. We know that G fibrations over X are classified by $[X, K(G,1)]$, homotopy classes of maps from X into the Eilenberg-Maclane space $K(G,1)$. On the other hand, $[X, K(G,1)]$ is canonically isomorphic to $H^1(X, G)$, the first cohomology group of X in G coefficients. We will define and establish similar relations between these notions in the category of real algebraic sets and entire rational maps.

Let X be a k dimensional compact nonsingular real algebraic set. Then the algebraic homology $H_*^A(X)$ of X is defined to be the subgroup of $H_*(X;\mathbb{Z}_2)$ whose elements are represented by algebraic subsets of X. The image of this subgroup under the Poincaré duality isomorphism is called the algebraic cohomology of X and denoted by $H_A^{k-*}(X)$ ([1]).

Let $\mathbb{F}$ denote the field $\mathbb{R}$ or $\mathbb{C}$. An algebraic $\mathbb{F}$-vector bundle ξ over a real algebraic set X consists of a real algebraic variety $E(\xi)$ and an entire rational map $\pi : E(\xi) \longrightarrow X$ such that for each $x \in X$, the fiber $\pi^{-1}(x)$ has the structure of an $\mathbb{F}$-vector space. Moreover, it is required that for each $x \in X$ there exist a Zariski neighborhood U of x, a nonnegative integer p and a commutative diagram

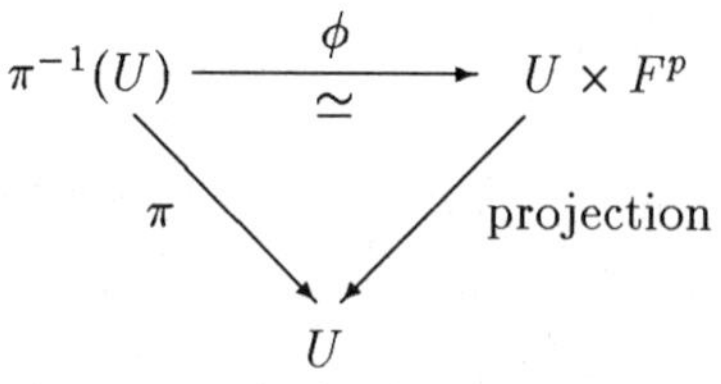

where ϕ is an isomorphism, which is $\mathbb{F}$-linear on every fiber. The notions of homomorphism, isomorphism, etc., of algebraic $\mathbb{F}$-vector bundles are defined in the obvious way.

An algebraic $\mathbb{F}$-vector bundle ξ over X is said to be *strongly algebraic* if ξ is algebraically isomorphic to an algebraic subbundle of a trivial bundle. An algebraic vector bundle over a real algebraic set is strongly algebraic if and only if the total space is a real algebraic set (p.94 in [4] or [11]).

EXAMPLE. For $p \leq n$ define

$$G_{n,p}(\mathbb{F}) = \{A \in M(n,\mathbb{F}) : A^2 = A = A^*, \quad \mathrm{tr}(A) = p\}$$

which is an algebraic model for the Grassmannian of p-dimensional $\mathbb{F}$-vector subspaces in $\mathbb{F}^n$. $G_{n,p}$ is a nonsingular real algebraic set. The canonical $\mathbb{F}$-vector bundle $\gamma_{n,p}(\mathbb{F}) = (E(\gamma_{n,p}(\mathbb{F})), \pi, G_{n,p}(\mathbb{F}))$ over $G_{n,p}(\mathbb{F})$ defined by

$$E(\gamma_{n,p}(\mathbb{F})) = \{(A,x) \in G_{n,p}(\mathbb{F}) \times \mathbb{F}^n : Ax = x\}$$

where $\pi(A, x) = A$, is strongly algebraic. More generally, if $f : X \longrightarrow G_{n,p}(\mathbb{F})$ is an entire rational map, then the pull back bundle $f^*\gamma_{n,p}(\mathbb{F})$ is strongly algebraic. In fact, each strongly algebraic vector bundle over X is of this form.

The Stiefel-Whitney classes of a strongly algebraic real vector bundle lie in the algebraic cohomology of X, $H_A^*(X)$. The Chern classes of a strongly algebraic complex vector bundle lie in the groups $H^*_{\mathbb{C}-Alg}(X, \mathbb{Z})$. For more detailed knowledge of algebraic vector bundles one may look at [5, 14].

We know that $\mathbb{R}P^\infty$ is a $K(\mathbb{Z}_2, 1)$ and therefore by the example in the remark below Theorem 2.2 $S^\infty \longrightarrow \mathbb{R}P^\infty$ is a universal algebraic $\mathbb{Z}_2$ fibration (see Definition 4.1). Let $[X, K(\mathbb{Z}_2, 1)]_A$ denote the subset of equivalence classes in $[X, K(\mathbb{Z}_2, 1)]$ which are represented by some entire rational maps from X to $K(\mathbb{Z}_2, 1) = \mathbb{R}P^\infty$.

PROPOSITION 3.1. *Let X be a compact real algebraic set. Then the natural isomorphism $\phi : [X, K(\mathbb{Z}_2, 1)] \longrightarrow H^1(X, \mathbb{Z}_2)$ induces a bijection*

$$\phi : [X, K(\mathbb{Z}_2, 1)]_A \longrightarrow H^1_A(X).$$

PROOF. First note that every element $w \in H^1(X, \mathbb{Z}_2)$ is the $w_1(L)$ for some line bundle $\pi : L \longrightarrow X$. Also $\mathbb{R}P^n = G_{n,1}(\mathbb{R})$. So we are done by the above preliminary facts. □

3.2. Odd order group case. Let G be a finite Abelian group of odd order. We will build a sequence of algebraic sets whose direct limit is a $K(G, 1)$. Since $K(G_1 \times G_2, 1) = K(G_1, 1) \times K(G_2, 1)$ we can assume that $G = \mathbb{Z}_r$ where $r = p^s$ is an (odd) prime power. Let G act on $(\mathbb{C}^l)^*$ by multiplication with $e^{2\pi i/r}$ on each component. Then the inclusion $(\mathbb{C}^l)^* \longrightarrow (\mathbb{C}^{l+1})^*$ is equivariant. The direct limit

$$\mathrm{L(r)}=\varinjlim\, (\mathbb{C}^l)^*/\mathrm{G}$$

is nothing but a thickened infinite Lens space which is a $K(G, 1)$. Note that each $(\mathbb{C}^l)^*$ is a nonsingular Zariski open real algebraic set and G acts on it freely. The Zariski open real algebraic set $(\mathbb{C}^l)^*$ is equivariantly isomorphic to the algebraic set

$$\{(x, t) \in \mathbb{R}^{2l} \times \mathbb{R} : |x|^2 \cdot t = 1\}$$

where the G acts only on the $\mathbb{R}^{2l}$ component (note that the defining equation $|x|^2 \cdot t = 1$ is G invariant). Applying part (c) of Theorem 2.1 we see that the quotients $(\mathbb{C}^l)^*/G$ are also nonsingular real algebraic sets.

REMARK. As we will need in Section 4.2, the quotient map $q : L(r) \longrightarrow \mathbb{C}P^\infty$ is entire rational. We will explain this on an example rather than giving the formal general argument: Assume that $G = \mathbb{Z}_3$ and consider its action on $(\mathbb{C}^2)^*$. Then the quotient $(\mathbb{C}^2)^*/G$ can be identified with the image of $F : (\mathbb{C}^2)^* \longrightarrow \mathbb{R}^m$ where

$$F(z_1, z_2) = (z_1^3, z_1^2 z_2, z_1 z_2^2, z_2^3, y_9, \ldots, y_m)$$

(let $y_1, \ldots, y_8$ be the real and imaginary parts of first 4 complex coordinates and we regard F as a map from $\mathbb{R}^4$ to $\mathbb{R}^m$; see Section 2). Then the quotient map $q : L(r) \longrightarrow \mathbb{C}P^\infty$ restricted to $(\mathbb{C}^2)^*/G$ is given by

$$q(w_1, w_2, w_3, w_4, t_9, \ldots, t_m) = [w_1, w_2, w_3, w_4]$$

which is clearly entire rational. Note that the composition $(q \circ F) : (\mathbb{C}^2)^* \longrightarrow \mathbb{C}P^4$ is given by

$$(z_1, z_2) \longrightarrow [z_1^3, z_1^2 z_2, z_1 z_2^2, z_2^3].$$

We know that $H^1(X, G) = [X, K(G, 1)] = [X, L(r)]$ and therefore one can define the following.

DEFINITION 3.2. The algebraic first cohomology group $H_A^1(X, G)$ of X with coefficients in G is defined to be the subgroup generated by the elements in $H^1(X, G)$ which have entire rational representatives in $[X, L(r)]$.

PROPOSITION 3.3. *Every element in $H_A^1(X, G)$ has an entire rational representative in $[X, L(r)]$.*

PROOF. For an Abelian group G and a positive integer n there is a unique H space structure (up to homotopy) on any $K(G, n)$ (Theorem 7.13 in Chap. 5 of [22]). Moreover, this H space structure determines a natural Abelian group structure on $[X, K(G, n)]$ so that the bijection $H^n(X, G) = [X, K(G, 1)]$ is a group isomorphism for any CW-complex X (Theorem 7.14 in Chap. 5 of [22]). In the next paragraph, we will show that $K(G, 1)$ has an H space structure whose multiplication and inverse operation are given by entire rational maps. So, if $a, b \in [X, L(r)]$ have entire rational representatives then $a - b$ will have an entire rational representative and therefore every element in $H_A^1(X, G)$ has an entire rational representative in $[X, L(r)]$.

Since the product of H spaces is another H space (with the coordinatewise multiplication and the coordinatewise inverse operation) it is enough to prove the proposition for the case $G = \mathbb{Z}_r$, where $r = p^j$ is a prime power. We can view the direct limit $(\mathbb{C}^\infty)^*$ of $(\mathbb{C}^n)^*$'s as the set of all nonzero complex polynomials by identifying $(a_0, a_1, a_2, \ldots, a_n, 0, 0, \ldots)$ with the complex polynomial $a_0 + a_1 z + a_2 z^2 + \cdots + a_n z^n$. Then the polynomial multiplication induces a multiplication on $(\mathbb{C}^\infty)^*$. Note that this is an entire rational operation. It descends to a multiplication in $L(r)$, $\rho : L(r) \times L(r) \longrightarrow L(r)$. The inverse operation $Inv : L(r) \longrightarrow L(r)$ in $L(r)$ is the one induced from the complex conjugation in $(\mathbb{C}^\infty)^*$. To see this, consider the composition of the map $id_{L(r)} \times Inv : L(r) \longrightarrow L(r) \times L(r)$ with the multiplication $\rho : L(r) \times L(r) \longrightarrow L(r)$ and see that this composition lifts to $(\mathbb{C}^\infty)^*$ and therefore is homotopic to a constant map.□

REMARKS.

1) The above constructions of $K(G, 1)$ does not work for Abelian 2-groups. The reason is that in this case the set W in Theorem 2.2 is not empty and therefore the geometric quotient is not equal to the algebraic quotient.

2) Let $G = (\mathbb{Z}_2)^k \times H$ and H an odd order Abelian group. Then, we define $H_A^1(X, G)$ as in the case of odd order Abelian G, where we take $K(H, 1) \times (\mathbb{R}P^\infty)^k$ as an algebraic model for $K(G, 1)$.

3) Exactly in the same way $\mathbb{C}P^\infty$ and $\mathbb{R}P^\infty$ are H spaces. In the latter case we consider real polynomials instead of complex ones. Note that $Inv : \mathbb{R}P^\infty \longrightarrow \mathbb{R}P^\infty$ is the identity map.

4. Main results: Algebraic principal G fibrations

4.1. Main Theorems. As in the previous section the phrase *G fibration* will actually mean *principal G fibration.*

DEFINITION 4.1. Let G be a finite group and X and Y be real algebraic sets. A G fibration $\pi : X \longrightarrow Y$ over the real algebraic sets Y is said to be algebraic if the G action on X is algebraic and the map $\pi : X \longrightarrow Y$ is the corresponding algebraic quotient map as in Section 2.

From this point on, G will be $(\mathbb{Z}_2)^k \times H$ for some $k > 0$ and an odd order Abelian group H (see Remark 1 at the end of Section 3). Let $\pi : M \longrightarrow Y$ be a smooth G fibration over the nonsingular real algebraic set Y. Our aim is to realize the smooth G fibration $\pi : M \longrightarrow Y$ algebraically. This is not always possible. Indeed, we will show that the quotient group $H^1(Y, G)/H^1_A(Y, G)$ carries the obstruction to do that.

In the case of $\mathbb{Z}_2$ fibrations over a compact nonsingular real algebraic set Y we have the following result. In the proof we make use of strongly algebraic line bundles introduced in Section 3.

THEOREM 4.2. *Let $\pi : M \longrightarrow Y$ be a smooth $\mathbb{Z}_2$ fibration over a compact nonsingular algebraic set Y. Then there exists an algebraic model $F : X \longrightarrow Y$ of the $\mathbb{Z}_2$ fibration $\pi : M \longrightarrow Y$ if and only if the first Stiefel Whitney class $w_1(L)$ of the line bundle L, associated to the $\mathbb{Z}_2$ fibration $\pi : M \longrightarrow Y$, lies in the first algebraic cohomology group $H^1_A(Y)$.*

PROOF. Assume that there exists an algebraic model of the fibration $\pi : M \longrightarrow Y$ as in Definition 4.1. Now $X \times \mathbb{R}$ is a nonsingular algebraic set and there is a free $\mathbb{Z}_2$ action on it: if $\sigma' : X \longrightarrow X$ is the $\mathbb{Z}_2$ action on X then we define the $\mathbb{Z}_2$ action on $X \times \mathbb{R}$ by $\sigma : X \times \mathbb{R} \longrightarrow X \times \mathbb{R}$ where $\sigma(x, t) = (\sigma'(x), -t)$. It is easy to see that for this $\mathbb{Z}_2$ action on $X \times \mathbb{R}$ the set W in Theorem 2.2 is empty and therefore the quotient, say L', is a nonsingular real algebraic set. $L' \longrightarrow Y$ is an algebraic line bundle over Y, isomorphic to $L \longrightarrow Y$, and its total space L' is an algebraic set. So $L' \longrightarrow Y$ is a strongly algebraic line bundle over Y (see p. 94 in [4] or [11]), and therefore $w_1(L) = w_1(L')$ is algebraic.

Conversely assume that $w_1(L)$ is algebraic. Then, L is C^0 isomorphic to a strongly algebraic line bundle L' over Y (see Proposition B in [14]). Since both bundles are smooth they are smoothly isomorphic. There is an entire rational map $\phi : Y \longrightarrow G(m, 1)$ for some m, so that L' is the pull back of the canonical line bundle $P : E(m, 1) \longrightarrow G(m, 1)$. Evidently, the total space S^{m-1} of the unit sphere bundle of $E(m, 1)$ is a nonsingular algebraic subset of $E(m, 1)$. Now, let X be the algebraic set defined by

$$X = \{(y, e) \in Y \times S^{m-1} : P(e) = \phi(y)\}.$$

By writing down the Jacobian of the locally defining ideal at any point (y, e) of X and noting that y and e are nonsingular points of Y and S^{m-1} respectively, we see that X is a nonsingular algebraic set. The inherited $\mathbb{Z}_2$ action on X is given by $(y, e) \longrightarrow (y, -e)$. By the construction X and M are equivariantly diffeomorphic to each other. On the other hand, the quotient $X/\mathbb{Z}_2$, which is equal to $X//\mathbb{Z}_2$ (see the remark following Theorem 2.2), is the graph of $\phi : Y \longrightarrow G(m,1)$, and therefore is isomorphic to Y. □

REMARK. Each element in $H^1(Y;\mathbb{Z}_2)$ is the $w_1(L)$ of some line bundle L over Y. Therefore, the existence of compact nonsingular real algebraic sets with $H^1(Y;\mathbb{Z}_2) \neq H^1_A(Y;\mathbb{Z}_2)$ ([1]) gives examples of smooth $\mathbb{Z}_2$ fibrations $M \longrightarrow Y$ which can not be realized algebraically.

For the case of odd order cyclic G we have a similar result.

THEOREM 4.3. *Let G be a finite cyclic group of odd order r, and $\pi : M \longrightarrow Y$ be a G fibration over a compact nonsingular real algebraic set Y. Then, this G fibration is algebraically realized if and only if it has an entire rational classifying map $\phi : Y \longrightarrow K(G,1)$; i.e., the homotopy class $[\phi]$ of $\phi : Y \longrightarrow K(G,1)$ is in $H^1_A(Y,G)$.*

PROOF. The sufficiency part is the same as the one in the previous theorem. Conversely suppose that $F : X \longrightarrow Y$ is an algebraic model of our G fibration. Then G acts on $X_{\mathbb{C}} \times \mathbb{C}$ (G acts on $X_{\mathbb{C}}$ already; the G action on $\mathbb{C}$ is given via multiplication with $e^{2\pi i/r}$). The ring of polynomial functions (complex) on $X_{\mathbb{C}} \times \mathbb{C}$ is equal to $S_{\mathbb{C}}[w]$ (one variable polynomial ring over $S_{\mathbb{C}}$, where S and $S_{\mathbb{C}}$ are as in Section 2). The invariant subring R^* of $S_{\mathbb{C}}[w]$ under the action of G evidently contains $R_{\mathbb{C}}$, the invariant subring of $S_{\mathbb{C}}$ under the action of G. If a polynomial in $S_{\mathbb{C}}[w]$ is an invariant polynomial then each summand of the polynomial is an invariant polynomial. Therefore, the invariant ring R^* is generated by the generators of $R_{\mathbb{C}}$ and some other elements of the form fw^j where $f \in S_{\mathbb{C}} - R_{\mathbb{C}}$. Let p be a prime number dividing r and $s \geq 1$ be the largest integer so that p^s divides r. Consider the subset

$$P = \{f_1 w^{j_1}, \cdots, f_l w^{j_l}\}$$

of this generating set with the properties:

i) None of the j's is divisible by p;

ii) For each $x \in X$, $f_k(x) \neq 0$ for some $k = 1, \ldots, l$.

Note that P is non empty, because otherwise the function whose coordinates are the generators of R^* would not separate the G orbits of $X_{\mathbb{C}} \times \mathbb{C}$. Let $fw^j \in P$ and $\rho \in G$ be a generator. Since, $fw^j = \rho(fw^j) = \rho(f)(e^{-2j\pi i/r}w^j)$ we see that $\rho(f) = e^{2j\pi i/r}f$. Since j is not divisible by p replacing f by some power of f, we can assume that $\rho(f) = e^{-2\pi i/p^s}f$. Now do the same thing for the other primes. Putting all these f's together we get a polynomial map $H : X \longrightarrow (\mathbb{C}^L)^*$ (L is the sum of l's) which embeds each fiber of $F : X \longrightarrow Y$ equivariantly. By the construction of H, the composition of H with the projection $(\mathbb{C}^L)^* \longrightarrow L(r)$ preserves the fibers of $F : X \longrightarrow Y$ and therefore induces a polynomial map $\phi : Y \longrightarrow L(r)$ classifying the fibration. □

This theorem generalizes to the groups $G = (\mathbb{Z}_2)^k \times H$ where H is an odd order Abelian group. However, first we need some preliminaries. Let M be a smooth manifold and $G = G_1 \times \cdots \times G_l$ be a finite group acting freely on M. For each $i = 1, \ldots, l$ let $\pi_i : M \longrightarrow M_i = M/(G/G_i)$ and $\theta_i : M_i \longrightarrow N$ be the smooth G/G_i and G_i fibrations respectively, where $N = M/G$. The fiber product

$$Z = \{(x_1, \ldots, x_l) : x_i \in M_i \text{ and } \theta_i(x_i) = \theta_j(x_j) \text{ for all } i,j = 1, \ldots, l\}$$

is clearly a smooth manifold and the two smooth fibrations $\pi : M \longrightarrow N$ and $\theta : Z \longrightarrow N$ are isomorphic:

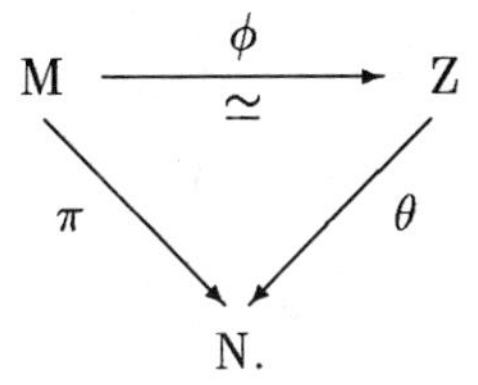

The above discussion holds also in the real algebraic category.

LEMMA 4.4. *Let X be a nonsingular real algebraic set and $G = G_1 \times \cdots \times G_l$ be a finite group acting freely and algebraically on X. Let $F_i : X \longrightarrow X_i = X/(G/G_i)$ and $\theta_i : X_i \longrightarrow Y$ be the real algebraic G/G_i and G_i fibrations respectively, where $Y = X/G$. Moreover, assume that all the topological quotients considered here are equal to the algebraic ones. Let Z be the fiber product*

$$Z = \{(x_1, \ldots, x_l) : x_i \in X_i \quad \text{and} \quad \theta_i(x_i) = \theta_j(x_j) \quad \text{for all} \quad i, j = 1, \ldots, l\}.$$

Then the real algebraic fibrations $F : X \longrightarrow Y$ and $\theta : Z \longrightarrow Y$ are algebraically isomorphic:

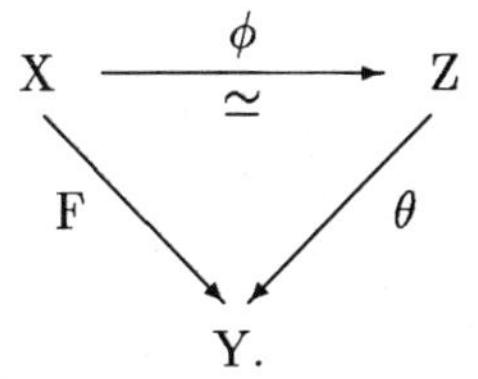

PROOF. Note that by the above paragraph it is enough to show that the map $\phi : X \longrightarrow Z$ is a real algebraic isomorphism. The algebraic map $\phi : X \longrightarrow Z$ defined by $x \mapsto (F_1(x), \ldots, F_l(x))$ is a diffeomorphism because the G action is free. Actually, this diffeomorphism extends to some Zariski open neighborhoods of X and Z in their complexifications. Therefore, the complexified map $\phi_{\mathbf{C}} : X_{\mathbf{C}} \longrightarrow Z_{\mathbf{C}}$ has degree one. Now by Proposition 3.17 in [15] the induced map $i : K(Z_{\mathbf{C}}) \longrightarrow K(X_{\mathbf{C}})$ from $\phi_{\mathbf{C}} : X_{\mathbf{C}} \longrightarrow Z_{\mathbf{C}}$ is an isomorphism where $K(Z_{\mathbf{C}})$ and $K(X_{\mathbf{C}})$ are the fields rational functions on $Z_{\mathbf{C}}$ and on $X_{\mathbf{C}}$ respectively. So the map $\phi_{\mathbf{C}} : X_{\mathbf{C}} \longrightarrow Z_{\mathbf{C}}$ is a birational isomorphism of these complex algebraic sets. However, since $\phi : X \longrightarrow Z$ is a diffeomorphism the birational inverse is well defined on Z. So $\phi : X \longrightarrow Z$ is a real algebraic isomorphism. □

THEOREM 4.5. *Let $G = (\mathbb{Z}_2)^k \times H$ where H is an odd order Abelian group and $\pi : M \longrightarrow Y$ be a smooth G fibration over a compact nonsingular real algebraic set Y. Then, this G fibration is algebraically realized if and only if it has an entire rational classifying map $\phi : Y \longrightarrow K(G, 1)$; i.e., the homotopy class $[\phi]$ of $\phi : Y \longrightarrow K(G, 1)$ is in $H^1_A(Y, G)$.*

PROOF. Write G as $G_1 \times \cdots \times G_l$ where each G_i for $i = 1, \ldots, l$ is a cyclic p_i group for some prime p_i or is a $\mathbb{Z}_2$. Then $K(G,1) = K(G_1,1) \times \cdots \times K(G_l,1)$. Now we are done by Lemma 4.4 and Theorems 4.2 and 4.3. □

4.2. The Bockstein Sequence and Examples. Let Y be a nonsingular real algebraic set. Consider the Bockstein cohomology exact sequence coming from the exact sequence of Abelian groups

$$0 \longrightarrow \mathbb{Z} \xrightarrow{\times r} \mathbb{Z} \longrightarrow \mathbb{Z}_r \longrightarrow 0$$

$$0 \longrightarrow H^1(Y,\mathbb{Z}) \xrightarrow{\times r} H^1(Y,\mathbb{Z}) \xrightarrow{i_*} H^1(Y,\mathbb{Z}_r) \xrightarrow{\beta} H^2(Y,\mathbb{Z}) \longrightarrow \cdots.$$

Note that $H^1(Y,\mathbb{Z}_r) = [Y, L(r)]$, $H^2(Y,\mathbb{Z}) = [Y, \mathbb{C}P^\infty]$ and the map

$$[Y, L(r)] \xrightarrow{\beta} [Y, \mathbb{C}P^\infty]$$

is given by post composing with the quotient map $L(r) \longrightarrow \mathbb{C}P^\infty$ which is entire rational by the remark above Definition 3.2. Hence for any algebraic $\mathbb{Z}_r$ fibration $X \longrightarrow Y$ there is a strongly algebraic complex line bundle $L \longrightarrow Y$ over Y with $c_1(L) = \beta(k) \in H^2_{\mathbb{C}-Alg}(Y,\mathbb{Z})$ (see Section 3), where $k \in [Y, L(r)]$ is a classifying map of the fibration $X \longrightarrow Y$.

PROPOSITION 4.6. *For any positive odd integer r, there exits an algebraic model Y of the compact 3 dimensional Lens space $L(r,q)$ $(q \leq r)$ with $H^1_A(Y,\mathbb{Z}_r) = 0$. In particular, for any $n > 0$ with $n|r$, Y admits no nontrivial algebraic $\mathbb{Z}_n$ fibrations.*

PROOF. By the example 7.4 in [5] there exists an algebraic model Y of $L(r,q)$ with $H^2_{\mathbb{C}-Alg}(Y,\mathbb{Z}) = 0$. Note that $H^1(Y,\mathbb{Z}) = 0$. Let $n > 0$ be any divisor of r. By the above discussion $H^1_A(Y,\mathbb{Z}_n)$ injects into $H^2_{\mathbb{C}-Alg}(Y,\mathbb{Z})$ and thus $H^1_A(Y,\mathbb{Z}_n) = 0$. Now by Theorem 4.3 Y admits no nontrivial algebraic $\mathbb{Z}_n$ fibrations. □

PROPOSITION 4.7. *Let Y be a compact nonsingular real algebraic set with $H^1(Y,\mathbb{Z}_2) = H^1_A(Y,\mathbb{Z}_2)$. Also assume that the Bockstein homomorphism*

$$H^1(Y,\mathbb{Z}_r) \xrightarrow{\beta} H^2(Y,\mathbb{Z})$$

is the zero map. Then for any positive odd integer r, $H^1(Y,\mathbb{Z}_r) = H^1_A(Y,\mathbb{Z}_r)$.

PROOF. Let $a \in H^1(Y,\mathbb{Z}_r)$. Since $\beta(a) = 0$ we see that $a = i_*(b)$ for some $b \in H^1(X,\mathbb{Z}) = [X, S^1]$ where the entire rational embedding $i : S^1 \longrightarrow L(r)$ represents the generator of $\pi_1(L(r))$ (note that $S^1 \subseteq \mathbb{C}^* \subseteq (\mathbb{C}^\infty)^*$ and $S^1//\mathbb{Z}_r \subseteq (\mathbb{C}^\infty)^*//\mathbb{Z}_r = L(r)$ is isomorphic to S^1). Now since $H^1(Y,\mathbb{Z}_2) = H^1_A(Y,\mathbb{Z}_2)$ we can assume that $b = [f]$ for some entire rational map $f : X \longrightarrow S^1$ (Theorem 1.4 in [8] or [13]). Finally, $a = [i \circ f]$ is clearly in $H^1_A(Y,\mathbb{Z}_r)$ and thus $H^1(Y,\mathbb{Z}_r) = H^1_A(Y,\mathbb{Z}_r)$.
□

We have the following immediate corollaries.

COROLLARY 4.8. *Let Y be a 1-dimensional, compact, connected and nonsingular real algebraic set. Then for any positive odd integer $r > 1$, $H^1_A(Y,\mathbb{Z}_r) = H^1(Y,\mathbb{Z}_r) = \mathbb{Z}_r$.*

COROLLARY 4.9. *Let Y be a compact nonsingular real algebraic surface with $H^1(Y,\mathbb{Z}_2) = H^1_A(Y,\mathbb{Z}_2)$ and $G = \mathbb{Z}_2^k \times H$ where H is an odd order Abelian group. Then $H^1(Y,G) = H^1_A(Y,G)$. In particular any smooth G fibration over Y is real algebraically realized.*

PROOF. It is enough to prove the corollary for the cyclic factors of G. By the assumption it is true for the $\mathbb{Z}_2$ factors. For the $\mathbb{Z}_r$ factors where r is an odd integer, the Bockstein homomorphism

$$H^1(Y,\mathbb{Z}_r) \xrightarrow{\beta} H^2(Y,\mathbb{Z})$$

is the zero map, because $H^1(Y,\mathbb{Z}_r)$ is an odd order finite group and $H^2(Y,\mathbb{Z})$ is either $\mathbb{Z}$ or $\mathbb{Z}_2$, depending on whether Y is orientable or not. So we are done by Proposition 4.7. □

REMARKS.

1) Note that, the assumption in Proposition 4.7 that the Bockstein homomorphism

$$H^1(Y,\mathbb{Z}_r) \xrightarrow{\beta} H^2(Y,\mathbb{Z})$$

is the zero map, can not be dropped. Indeed, for the Lens space Y in Proposition 4.6, for which the Bockstein homomorphism is not trivial, we have $H^k(Y,\mathbb{Z}_2) = H^k_A(Y,\mathbb{Z}_2)$ for all k but $H^1_A(Y,\mathbb{Z}_r) = 0 \neq \mathbb{Z}_r = H^1(Y,\mathbb{Z}_r)$.

2) Another interesting question along these lines is: Given a closed smooth manifold M and a positive odd integer $r > 1$, does there exist an algebraic model X of M with $H^1(X,\mathbb{Z}_r) = H^1_A(X,\mathbb{Z}_r)$? If the Bockstein homomorphism

$$H^1(M,\mathbb{Z}_r) \xrightarrow{\beta} H^2(M,\mathbb{Z})$$

of M is zero as in Proposition 4.7 then M has such an algebraic model provided that M is not an unorientable surface. Indeed, we know that in this case there exists an algebraic model X of M with $H^1(X,\mathbb{Z}_2) = H^1_A(X,\mathbb{Z}_2)$ (see [7]). Now by Proposition 4.7 X has the desired property.

4.3. Algebraic Cohomology of Algebraic Quotients. Let G be a finite group acting algebraically on a compact algebraic set X such that $X/G = X//G$. Let $\pi : X \longrightarrow Y = X/G$ be the quotient map. Let p be prime number which is not a factor of the order $|G|$ of G. Then it is well known that $H^*(Y,\mathbb{Z}_p) = (H^*(X,\mathbb{Z}_p))^G$ ([9]). For algebraic cohomology groups we have the following analogous result.

PROPOSITION 4.10. *Let X, Y, G and p be as above. Then $H^1_A(Y,\mathbb{Z}_p)$ is isomorphic to $(H^1_A(X,\mathbb{Z}_p))^G$. Moreover, if $p = 2$ then $H^k_A(Y)$ is isomorphic to $(H^k_A(X))^G$ for all k.*

PROOF. Since $\pi^* : H^1(Y,\mathbb{Z}_p) \longrightarrow (H^1(X,\mathbb{Z}_p))^G$ is an isomorphism and the quotient map $\pi : X \longrightarrow Y = X/G$ is algebraic we see that $\pi^* : H^1_A(Y,\mathbb{Z}_p) \longrightarrow (H^1_A(X,\mathbb{Z}_p))^G$ is an injection. To show that it is onto let $\alpha \in (H^1_A(X,\mathbb{Z}_p))^G$. Then by Proposition 3.3 there exists an entire rational map $f : X \longrightarrow L(p)$ so that $f \circ g$ is homotopic to f for all $g \in G$. Since p and $|G|$ are relatively prime there exists an integer k so that F^k represents α, where F is the product of all $f \circ g$'s for $g \in G$. By construction F^k is G invariant and therefore it descends to a map $\widetilde{F^k} : Y \longrightarrow L(p)$. Hence we are done in this case.

For the second statement, just remember that algebraic $\mathbb{Z}_2$ homology classes are represented by algebraic subsets and then use the averaging idea in the above paragraph. Here instead of taking the products of the maps representing the cohomology classes we simply take the union of the algebraic subsets representing the homology classes. □

5. Algebraic G fibrations are not unique

Next two propositions will show that both the trivial and the nontrivial $\mathbb{Z}_2$ fibrations over S^1, the standard unit circle in $\mathbb{R}^2$, have many birationally different algebraic realizations. Let E_λ denote the nonsingular complex elliptic curve whose $j-invariant$ ([12]) is

$$j(\lambda) = 2^8(\lambda^2 - \lambda + 1)^3/\lambda^2(\lambda - 1)^2.$$

PROPOSITION 5.1. *S^1 has a one-parameter family of topologically trivial algebraic $\mathbb{Z}_2$ fibrations $X_a \longrightarrow S^1$ where $a > 2$ and each X_a is the real part of the nonsingular elliptic curve E_λ, $\lambda = (2-a)/4$.*

PROOF. Let a be a real number greater than 2. Let $a(x)$ be the polynomial $4x^2 + a - 2$. Consider the one dimensional nonsingular real algebraic set $X \subseteq \mathbb{R}^4$ given by

$$X_a = \{(x, y, u, v) \in \mathbb{R}^4 : x^2 + y^2 = 1, u^2 = y^2 a(x), v^2 = x^2 a(x), uv = xya(x)\}.$$

$\mathbb{Z}_2$ acts on X_a by sending (x, y, u, v) to $(x, y, -u, -v)$ and this gives a quotient map onto its image Y:

$$\pi : X_a \longrightarrow Y \quad \text{by} \quad \pi(x, y, u, v) = (x, y, u^2, v^2, uv).$$

Note that Y is isomorphic to S^1. So we have an algebraic $\mathbb{Z}_2$ fibration over S^1. This fibration has a section by $(x, y) \longrightarrow (x, y, y(a(x))^{1/2}, x(a(x))^{1/2})$ and therefore it is topologically trivial.

Next we show that the complexification of X_a has the properties stated in the Proposition. By Proposition 2.5 in [7], we can assume that X_a sits in some $\mathbb{R}P^n$, so that, its complexification in $\mathbb{C}P^n$ is nonsingular. The function field K_a of the complexification of X_a is $\mathbb{C}(x, y, w)$ modulo the relations $x^2+y^2 = 1, w^2 = a(x)$. By letting $s = x + iy$ and $t = w(x + iy)$ we get $K_a = \mathbb{C}(s, t)/(s^4 + as^2 + 1 - t^2)$. These function fields correspond to the well-known family of elliptic curves, the Jacobi Family (pp. 87-88 in [12]), with the j-invariant as above. □

PROPOSITION 5.2. *For each positive odd integer l there exists a topologically nontrivial algebraic $\mathbb{Z}_2$ fibration over S^1, $X_l \longrightarrow S^1$ where X_l is the real part of a nonsingular complex algebraic curve of genus at least $l - 1$.*

PROOF. Let a be a real number greater than 2. Let $f(z) = 1 + az^l + z^{2l}$ and consider the ramified double covering $\pi : Z \longrightarrow \mathbb{C}$ where

$$Z = \{(z, w) \in \mathbb{C}^2 : w^2 = f(z)\}$$

which is a complex algebraic curve and $\pi : Z \longrightarrow \mathbb{C}$ is the projection onto the first factor. Let $X_l \subset Z$ be the preimage of $S^1 \subset \mathbb{C}$ under $\pi : Z \longrightarrow \mathbb{C}$. It is easy to see that X_l is a 1-dimensional nonsingular real algebraic set. Note that $f(z)$ has $2l$ distinct roots so that l of them are inside and l of them are outside the unit circle S^1. Since l is an odd integer X_l is connected (to see this, note that, X_l is the pull back of the double covering $\mathbb{C}^* \longrightarrow \mathbb{C}^*$, $z \mapsto z^2$, by the map $f : S^1 \longrightarrow \mathbb{C}^*$ whose degree is an odd integer l). Write $z = x + iy$ and $w = u + iv$. Note that, on S^1 we can identify x and y with $\cos\theta$ and $\sin\theta$. Then,

$$X_l = \{(x, y, u, v) \in \mathbb{R}^4 : x^2 + y^2 = 1, u^2 = (1 + \cos l\theta)|f|/2, \\ v^2 = (1 - \cos l\theta)|f|/2, uv = \sin l\theta |f|\}$$

where $|f| = a + 2\cos l\theta$ is the norm of $f(z)$ on S^1. As in the above proof X_l is an algebraic $\mathbb{Z}_2$ fibration over S^1.

Let us consider the complexification of X_l. Let $g(x)$ be the unique polynomial, so that, $g(\cos\theta) = \cos l\theta$ (to get $g(x)$ write $z^l/2+\bar{z}^l/2$ for $\cos l\theta$). Note that $g(1) = 1$ and $g(-1) = -1$. Throw out the double roots of $1 + g(x)$ and denote the remaining polynomial by $h(x)$. For generic a, the polynomial $H(x) = (a+2g(x))h(x)$ is square free over $\mathbb{C}$. Also one can check that -1 is a simple root of $1+g(x)$ and therefore of $H(x)$ (to see this write $1 + z^l/2 + \bar{z}^l/2$ for $1 + \cos l\theta$ and $1 + z/2 + \bar{z}/2$ for $1 + \cos\theta$ where $z \in S^1$ and then divide the first one with the second).

The complex algebraic set W in $\mathbb{C}^3$ given by the equations $x^2 + y^2 = 1$ and $w^2 = H(x)$ is birationally isomorphic to the complexification of X_l. The point $(x, y, w) = (-1, 0, 0)$ is the only singular point of W. Let Y denote the quadratic in $\mathbb{C}^2$ given by $x^2 + y^2 = 1$. W is a double covering of Y ramified at the points $(x, y) \in Y$ with $H(x) = 0$. $(-1, 0, 0)$ is one of the ramification points. Since 1 and -1 are not roots of $a + 2g(x)$ we get at least $2l$ more ramification points of the fibrations. By blowing up the singular point $(-1, 0, 0)$ we get a nonsingular double covering with at least $2l$ ramification points. When we projectivize and desingularize the covering there will be still at least $2l$ ramification points. Call this new covering W'. Since the projectivization of Y is $\mathbb{C}P^1$, W' should have genus at least $l-1$. By Proposition 2.5 in [7], X_l has an embedding into some real projective space so that its complexification is nonsingular in the corresponding complex projective space and since it is isomorphic to W' it has genus at least $l - 1$. □

References

[1] Akbulut, S., King, H.: *Topology of real algebraic sets*, M.S.R.I. book series, Springer, New York Berlin Heidelberg, 1992.

[2] ———: *On approximating submanifolds by algebraic sets and a solution to the Nash conjecture*, Invent. Math. **107** (1992), 87-89.

[3] ———: *Algebraicity of immersions*, Topology **31**, no.4 (1992), 701-712.

[4] Benedetti, R., Tognoli, A.: *On real algebraic vector bundles*, Bull. Sci. Math. (2) **104** (1980), 89-112.

[5] Bochnak, J., Buchner, M., Kucharz, W.: *Vector bundles over real algebraic varieties*, K-Theory Journal (1988).

[6] Bochnak, J., Coste, M., Roy, M.F.: *Géometrie Algébrique Réelle*, Ergebnisse der Math. vol. 12, Springer, 1987.

[7] Bochnak, J., Kucharz, W.: *Algebraic models of smooth manifolds*, Invent. Math. **97** (1989), 585-611.

[8] ———: *Algebraic approximation of mappings into spheres*, Michigan Math. J. **34** (1987), 119-125.

[9] Bredon, G.: *Introduction to compact transformation groups*, Academic Press , New York, 1972.

[10] Dovermann, K.H., Masuda, M.: *Algebraic realization of manifolds with group actions*, (to appear in Advances in Mathematics).

[11] Huisman, J.: *On real algebraic vector bundles*, University of Utrecht publications **834** TA Utrecht, The Netherlands (1993).

[12] Husemoller, D.: *Elliptic curves*, Springer, New York Berlin Heidelberg, 1986.

[13] Ivanov, N.: *Approximation of smooth manifolds by real algebraic sets*, Russian Math. Surveys **37** (1982), 1-59.

[14] Kucharz, W.: *Vector bundles over real algebraic surfaces and threefolds*, Compositio Math. **60**, (1986), 209-225.

[15] Mumford, D.: *Algebraic geometry I, Complex projective varieties*, Springer, Berlin Heidelberg New York, 1976.

[16] Nash, J.: *Real algebraic manifolds*, Annals of Math. **56** (1952), 405-421.

[17] Ozan, Y.: Ph.D thesis. Michigan State University, 1994.

[18] Procesi,C., Schwarz,G.: *Inequalities defining orbit spaces*, Invent. Math. **81** (1985), 539-554.

[19] Schwarz, G.: *Smooth functions invariant under the action of a compact lie group*, Topology **14** (1975), 63-68.

[20] Seifert, H.: *Algebraische approximation von mannigfaltigkeiten*, Math. Zeitschrift. **41** (1936), 1-17.

[21] Tognoli, A.: *Su una Congettur di Nash*, Ann. Scuola Norm. Sup. Pisa **27** (1973), 167-185.

[22] Whitehead, G.W.: *Elements of homotopy theory*, Springer, New York Heidelberg Berlin, 1978.

ACKNOWLEDGMENT: The author would like to thank Professor Selman Akbulut for his encouragement throughout this work.

DEPARTMENT OF MATHEMATICS, MICHIGAN STATE UNIVERSITY, EAST LANSING, MI 48824.

E-mail address: ozan@math.msu.edu

Contemporary Mathematics
Volume **182**, 1995

Resolving Singularities of Maps

HENRY C. KING

May 20, 1994

ABSTRACT. A conjecture is given for resolution of singularities of analytic and algebraic functions. This conjecture can be proven in low dimensions. This conjecture is related to my work with Akbulut on topological characterization of real algebraic sets, and currently gives topological characterization in dimensions 4 and less. There is also a very beautiful application to the problem of factoring birational maps into blowups and their inverses.

1. How nice can you make a map?

I would like to talk about the following problem: what do maps look like? As is, this problem is much too general, so I will narrow the focus to something more specific, namely analytic maps. So the problem is: suppose $f\colon X \to Y$ is an analytic map between (real or complex) analytic spaces, then what does f look like. Suppose you wish to examine f near a point of X or Y, and in fact wish to see infinitesimal information. Imagine you had a microscope with variable power, in fact with power inversely proportional to the distance from your point (or submanifold) of interest so that the magnification became infinite at the point (or submanifold) in question. You could then actually see some infinitesimal information. The mathematical device which performs this microscopic function is blowing up. The blow up of a space X with center a manifold C is a map $\pi\colon X' \to X$ where X' is obtained from X by replacing each point of C with the directions in X which are normal to C.

I will now make the problem more precise. Given $f\colon X \to Y$, an analytic map between analytic spaces, can you perform a sequence of blowups of X, $\rho\colon X' \to X$ and a sequence of blowups of Y, $\pi\colon Y' \to Y$ so that there is an analytic lifting $g\colon X' \to Y'$ with $\pi g = f\rho$ and so that g is nice.

1991 *Mathematics Subject Classification.* Primary 14E15, 32S45; Secondary 14E05, 14E35, 14E40, 14P25, 32H99.

This paper is in final form and no version of it will be submitted for publication elsewhere.

Okay, so what am I up to, and what does 'nice' mean. In fact this question has arisen in two different contexts that I am aware of, both ending up with essentially the same idea of what 'nice' should mean. One context is in work of T. C. Kuo, which I have no details on, and another is in work of Akbulut and mine on topologically classifying algebraic varieties.

I will take a little time to explain this work with Akbulut, since it is dear to my heart. Suppose we have a real or complex algebraic variety X and we wish to understand its topology. Hironaka's theorem on resolution of singularities says that there is a sequence of blowups $\pi\colon X' \to X$ so that X' is a manifold and π is proper. So, for example in the complex case (with X irreducible), X is a quotient space of a manifold. But this doesn't help all that much in understanding the topology of X unless you have some idea of the collapsing which the map π does. However, if you have a good theorem on resolution of singularities of maps, you can assume that the collapsing maps are of a nice enough form that you get a good idea of the topology of X. So, for example, Akbulut and I can give a topological classification of real algebraic varieties of dimension less than or equal to four because we can do the above resolution of singularities of maps in low dimensions, see [**1**].

I mentioned resolution of singularities above, so let me state this theorem in a useful form. It was originally proven in [**3**]. This proof is extremely difficult but now there is a much more understandable proof by Bierstone-Milman.

THEOREM 1.1. *(resolution of singularities) Suppose that X is a real or complex analytic variety and $Z \subset X$ is a subvariety which contains the singular points of X, (thus $X - Z$ is an analytic manifold). Then there is a sequence of blowups $\pi\colon X' \to X$, whose centers all lie over Z, so that X' is nonsingular and so $\pi^{-1}(Z)$ is a union of divisors with normal crossings, i.e., $\pi^{-1}(Z)$ looks locally like a union of coordinate hyperplanes. If X is algebraic then all the centers can be chosen to be algebraic, so X' is an algebraic variety.*

DEFINITION 1.1. *As a shorthand, we will call a union of divisors with normal crossings a* tico, *short for 'transversely intersecting codimension one'. We sometimes also use the term* regular tico *for this since the definition of tico in* [**1**] *allows submersed manifolds also.*

DEFINITION 1.2. Tico coordinates *will mean local coordinates in which the tico is a union of coordinate hyperplanes.*

2. Submersive maps

Well, enough motivational digression, lets get on to what 'nice' should mean. One of the nicest kinds of maps I know are submersions. But we have no hope of blowing up to make g a submersion as the following example shows. Consider $f\colon \mathbb{C} \to \mathbb{C}$ given by $f(z) = z^2$. Any blowing up of $\mathbb{C}$ is just the identity, so we are stuck with the critical point at 0. So we cannot hope to blow up to make

f a submersion, but as it will turn out, we can hope to blow up to make any analytic map a combination of submersion and monomial map.

For example, suppose that we have a proper analytic map $f\colon X \to Y$ between complex varieties where Y is one dimensional. By resolution of singularities we may as well assume that X and Y are manifolds. By Sard's theorem and Remmert's proper mapping theorem, the critical values of f form a zero dimensional analytic subset B, and by resolution of singularities, we may assume that $f^{-1}(B)$ is a tico. We can do no better than this. But let us see what we have.

The map f restricts to a submersion on $X - f^{-1}(B)$. In the complex case (and in the real case also if you exercise a certain amount of care) the fact that $f^{-1}(B)$ is a tico implies that near any point of $f^{-1}(B)$ you can take local coordinates so that the map f is a monomial (see Lemma 6.2.1 of [**1**]).

So this result I just proved on resolution of singularities of maps where Y has dimension one gives some hint as to what 'nice' should mean. It is some sort of combination of submersion and monomial map. I will now give two equivalent precise definitions. I will also abandon the vague word 'nice' in favor of the word 'submersive'.

Suppose we have an analytic map $f\colon X \to Y$. We also suppose we have ticos $\mathcal{A}$ in X and $\mathcal{B}$ in X. We also suppose that f is a tico map, in the complex case this only means that $f^{-1}(|\mathcal{B}|) \subset |\mathcal{A}|$, in the real case, see [**1**] for a definition. Note that the ticos $\mathcal{A}$ and $\mathcal{B}$ give stratifications of X and Y, i.e., they break them up into a disjoint union of submanifolds of various dimensions. The codimension k stratum of X would be those points which are in exactly k divisors of $\mathcal{A}$. In other words, in tico coordinates the point is the intersection of exactly k hyperplanes.

We will show that the following two definitions are equivalent.

DEFINITION 2.1. *The map f is* submersive *if there are everywhere tico coordinates on X and Y so that f is a monomial map and furthermore, each connected component of a stratum of X is submersed to a stratum of Y.*

DEFINITION 2.2. *The map f is* submersive *if the logarithmic Jacobian of f has maximal rank everywhere.*

3. The logarithmic Jacobian

Hey, where did this logarithmic Jacobian come from? Take local tico coordinates in X and Y. Take logarithms of those coordinates in X and Y which define the ticos A and B. Take the Jacobian matrix of the resulting map. It is only defined on $X - A$, but in fact it will extend to A. This is the logarithmic Jacobian.

For example, consider $f\colon \mathbb{C}^3 \to \mathbb{C}^3$ given by $f(x,y,z) = (x^2y^3, xy^4e^x, z+xy)$. Let the ticos in both copies of $\mathbb{C}^3$ be given by $xy = 0$. Then letting $u = \ln x$ and $v = \ln y$ our map becomes $(u,v,z) \mapsto (2u+3v, u+4v+e^u, z+e^{u+v})$ which has

Jacobian matrix

$$\begin{pmatrix} 2 & 3 & 0 \\ 1+x & 4 & 0 \\ xy & xy & 1 \end{pmatrix}$$

So if one deleted the affine hyperplane $x = 5/3$ the map f would be submersive.

Using Definition 2.1, we can see it is submersive when $x \neq 5/3$ by making the coordinate change $r = xe^{-.6x}$, $s = ye^{.4x}$, $t = z + xy$. Then $f(r,s,t) = (r^2s^3, rs^4, t)$, so f is a pure monomial map and it submerses each stratum of positive codimension to the codimension 2 stratum $x = y = 0$.

Note that the logarithmic Jacobian depends on the tico coordinates chosen and I do not see it as any kind of natural linear transformation, just a matrix. However, one can extract invariants of the matrix such as its rank which do not depend on the choice of tico coordinates. Later we will also find useful the ideal generated by its $k \times k$ minor determinants. The reason the rank and these ideals are well defined is that changing tico coordinates changes the logarithmic Jacobian by multiplying by the logarithmic Jacobian of the coordinate change map, which is easily seen to be nonsingular.

4. Blowing up a map to make it submersive

The work Akbulut and I did leads naturally to the following conjecture, which if proven in a reasonable technical form will imply a topological characterization of real algebraic varieties in all dimensions. The term *critical point* below means a point of X which is either a singular point of X or maps to a singular point of Y or is nonsingular and maps to a nonsingular point, but the derivative map df has rank less than the dimension of Y.

CONJECTURE 4.1. *Let $f\colon X \to Y$ be a proper analytic map between real or complex analytic varieties. Suppose the critical points of f are nowhere dense in X. Then there are sequences of blowups $\pi\colon Y' \to Y$ and $\rho\colon X' \to X$, ticos $\mathcal{A}$ and $\mathcal{B}$ in X' and Y', and an analytic map $g\colon (X', \mathcal{A}) \to (Y', \mathcal{B})$ so that $\pi \circ g = f \circ \rho$ and g is submersive. In the algebraic case (X and Y algebraic varieties, f regular) then g is regular and the blowups are algebraic.*

THEOREM 4.1. *The above conjecture is true if* $\dim Y \leq 1$ *or if* $\dim X \leq 3$.

The case $\dim Y = 0$ is trivial and the proof for $\dim Y = 1$ given above in the complex analytic case works in the real or algebraic setting also. The case $\dim X = \dim Y = 2$ is contained in the proof of Theorem 7.4.1 of [**1**]. The cases $\dim X = 3$ and $\dim Y = 2$ or 3 are considerably harder. The proof will appear elsewhere but I will outline the main ideas later in this paper.

When Akbulut and I first realized we should prove the above conjecture, we despaired of ever doing so since it is more difficult than resolution of singularities of varieties and Hironaka's proof is quite delicate, not appearing to lend itself to modification for other ends. That is why we found Bierstone's and Milman's simple proof of resolution of singularities so exciting. Finally, an ordinary mathematician can understand resolution of singularities. Moreover, since the proof uses a locally defined complexity, it can be adapted for other uses.

5. A simplicial complex associated to a tico

Suppose $(X, \mathcal{A})$ is a regular tico. We may associate to it a simplicial complex $\mathcal{K}(X, \mathcal{A})$ as follows. The vertices of $\mathcal{K}(X, \mathcal{A})$ are the nonempty sheets (i.e., divisors) of $\mathcal{A}$. There is a k-simplex with vertices $A_0, A_1, \ldots, A_k$ if the intersection $\bigcap_{i=0}^k A_i$ is nonempty. We let $|\mathcal{K}(X, \mathcal{A})|$ denote the underlying topological space of this complex.

Let $f\colon (X, \mathcal{A}) \to (Y, \mathcal{B})$ be a tico map with constant exponents. The notion of *constant exponents* is defined in [**1**], but a tico map will automatically have constant exponents if all the divisors in $\mathcal{A}$ and $\mathcal{B}$ are irreducible. We suppose also that f is *exact* which means that $f^{-1}(|\mathcal{B}|) = |\mathcal{A}|$. Then there is an induced map

$$\mathcal{K}(f)\colon |\mathcal{K}(X, \mathcal{A})| \to |\mathcal{K}(Y, \mathcal{B})|$$

which satisfies the usual functorial properties $\mathcal{K}(f \circ g) = \mathcal{K}(f) \circ \mathcal{K}(g)$ and $\mathcal{K}(id) = id$. This map is not piecewise linear, it involves radial projection which is not piecewise linear. This map is determined by the exponents of the map f, i.e., when you look at the map in local coordinates it is a monomial and the constant exponent hypothesis means the exponents of this monomial depend only on the sheets involved, not the point chosen.

Suppose $\pi\colon (X', \mathcal{A}') \to (X, \mathcal{A})$ is a tico blowup with nonempty center $C = \bigcap_{i=0}^k A_i$ for some sheets $A_i \in \mathcal{A}$. (Here *tico blowup* just means that the sheets in $\mathcal{A}'$ consist of the strict transforms of the sheets of $\mathcal{A}$ and also the exceptional divisor $\pi^{-1}(C)$.) Then the map $\mathcal{K}(\pi)$ is a homeomorphism and in fact gives a subdivision of $\mathcal{K}(X, \mathcal{A})$ as follows. Let σ be the k-simplex of $\mathcal{K}(X, \mathcal{A})$ with vertices $A_0, \ldots, A_k$. Then $\mathcal{K}(X', \mathcal{A}')$ is obtained from $\mathcal{K}(X, \mathcal{A})$ by replacing the closed star of σ with the cone on its boundary. The map $\mathcal{K}(\pi)$ takes the new vertex, the exceptional divisor, to the barycenter of σ.

For example, consider a manifold X with a tico $\mathcal{A}$ consisting of 4 sheets A_1, A_2, A_3, A_4. We are given that $A_1 \cap A_2 \cap A_3$ and $A_1 \cap A_2 \cap A_4$ are nonempty, but $A_1 \cap A_2 \cap A_3 \cap A_4$ is empty. Then $\mathcal{K}(X, \mathcal{A})$ consists of two triangles joined along the side $A_1 A_2$.

We will perform two sequences of blowups. For the first sequence we blow up $A_1 \cap A_2$, after which we blow up the intersection of the strict transform of A_3 with the exceptional divisor. For the second sequence, we first blow up $A_1 \cap A_2 \cap A_3$ after which we blow up the intersection of the strict transforms of A_1 and A_2.

Figure 1 indicates the changes to the associated simplicial complex as we perform the indicated blowups. In the first sequence, we blow up $A_1 \cap A_2$. The resulting space has an exceptional divisor, the inverse image of $A_1 \cap A_2$ which we denote A_5, the corresponding vertex in the complex being labeled 5 in the

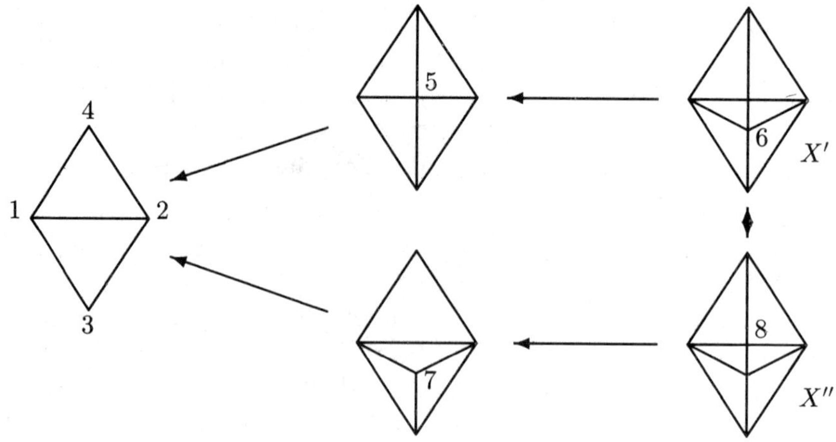

FIGURE 1. $\mathcal{K}(X, \mathcal{A})$ example

figure. We then blow up the intersection of A_5 and the strict transform of A_3, giving a new exceptional divisor A'_6. Call the resulting space X'. In X' is a tico $\mathcal{A}' = \{A'_1, A'_2, \dots, A'_6\}$ where A'_i is the strict transform of A_i. For the second sequence, we blow up $A_1 \cap A_2 \cap A_3$ and denote the resulting exceptional divisor A_7. We then blow up the intersection of the strict transforms of A_1 and A_2, with the resulting divisor A''_8. Call the resulting space X''. In X'' is a tico $\mathcal{A}'' = \{A''_1, A''_2, \dots, A''_4, A''_7, A''_8\}$ where A'_i is the strict transform of A_i.

In looking at the complex $\mathcal{K}(X', \mathcal{A}')$ in Figure 1, the reader may think I am cheating at first, since A_6 was supposed to appear at the barycenter of the simplex between 5 and 3, it looks in the figure as if it is 1/3 the way along this simplex. But this is right, since the the map identifying this subdivision with our original simplicial complex $\mathcal{K}(X, \mathcal{A})$ is not piecewise linear. This is the source of the apparent distortion. More precisely, A_5 is mapped to the point with projective barycentric coordinates $(1 : 1 : 0)$ which is the barycenter of A_1A_2, but A_6 is mapped to the point with projective barycentric coordinates $(1 : 1 : 1)$ since $(1, 1, 0) + (0, 0, 1) = (1, 1, 1)$. This point is not on the center of the line from $(0 : 0 : 1)$ to $(1 : 1 : 0)$ but is 1/3 the way along, and actually coincides with the barycenter of the triangle $A_1A_2A_3$.

I should tell the reader what projective barycentric coordinates are. A point in a simplex with projective barycentric coordinates $(a_1 : a_2 : \dots : a_n)$ has barycentric coordinates $(a_1/a, a_2/a, \dots, a_n/a)$ where $a = \sum_{i=1}^n a_i$. As we will see below, it is often convenient to describe a point's projective barycentric coordinates. This could have been avoided by defining $\mathcal{K}(X, \mathcal{A})$ differently, say taking the open cone on the $\mathcal{K}(X, \mathcal{A})$ I have defined here. Then $\mathcal{K}(f)$ would be PL, and you would not need exactness of f to define $\mathcal{K}(f)$. But I find it preferable to define things as I have here and use projective barycentric coordinates.

Note that we end up with the same complex at the end of each of the two sequences of blowups. In fact the subdivision of $\mathcal{K}(X, \mathcal{A})$ is exactly the same

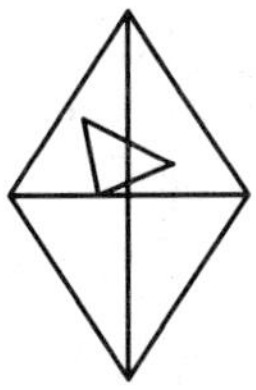

FIGURE 2. Example of map which doesn't lift

subdivision. As we shall see, this means that the resulting ticos are the same. There is an isomorphism $h\colon X' \to X''$ so that $h(A_i') = A_i''$ for $i = 1, 2, 3, 4$ and $h(A_5') = A_8''$ and $h(A_6') = A_7''$. This follows from the following discussion.

First of all, suppose $\pi\colon (X', \mathcal{A}') \to (X, \mathcal{A})$ is a sequence of tico blowups so that all centers are intersections of sheets. As we said above, the map $\mathcal{K}(\pi)\colon |\mathcal{K}(X', \mathcal{A}')| \to |\mathcal{K}(X, \mathcal{A})|$ is a homeomorphism and in fact gives a subdivision of $|\mathcal{K}(X, \mathcal{A})|$. Now suppose that $f\colon (Y, \mathcal{B}) \to (X, \mathcal{A})$ is an exact tico map with constant exponents. It is useful to know if f lifts to a map to X'. The amazing result is that f lifts to a map $g\colon Y \to X'$ with $f = \pi g$ if and only if the map $\mathcal{K}(f)$ has the the property that for each simplex σ of $\mathcal{K}(Y, \mathcal{B})$ there is a simplex δ of $\mathcal{K}(X', \mathcal{A}')$ so that $\mathcal{K}(f)(\sigma) \subset \mathcal{K}(\pi)(\delta)$. In other words, $\mathcal{K}(f)$ maps nicely into the resulting subdivision. Furthermore, if the lifting g exists then it is a tico map. For example, Figure 2 shows an example of a map which doesn't lift. The blowup is the first blowup with center $A_1 \cap A_2$ in the above example. The skewed triangle is the image of some simplex. Since it is not contained in a simplex of the subdivision, the map does not lift.

Going back to our example, we see that the map $\pi'\colon (X', \mathcal{A}') \to (X, \mathcal{A})$ lifts to a map $h\colon (X', \mathcal{A}') \to (X'', \mathcal{A}'')$ and $\pi''\colon (X'', \mathcal{A}'') \to (X, \mathcal{A})$ lifts to a map $h'\colon (X'', \mathcal{A}'') \to (X', \mathcal{A}')$. Since hh' and $h'h$ must be the identity on dense subsets (the complement of the inverse images of the centers) they give an isomorphism.

6. Factoring birational maps

This simplicial complex assigned to a tico seems useful in investigating the problem of factoring birational maps. The reason is that if f is birational and submersive and a few mild assumptions are satisfied, then $\mathcal{K}(f)$ is a homeomorphism (Theorem 7.6). What is more, there are strong conditions on this homeomorphism given by Theorem 7.7, the matrix formed by the coordinates of the images of the vertices of a simplex has determinant ± 1.

So suppose you have a birational map $f\colon X \to Y$ between complex algebraic varieties and wish to prove that it is a composition of blowup maps and their inverses, or, who knows, find obstructions to its having such a decomposition. As a first step, assuming you know the map resolution conjecture Theorem 4.1, you may as well assume that there are ticos $\mathcal{A}$ and $\mathcal{B}$ so that $f\colon (X, \mathcal{A}) \to (Y, \mathcal{B})$ is an exact submersive tico map. At that point you try to see if f is a composition of tico blowups and their inverses. But here is where the simplicial complex associated to a tico is useful. The map $\mathcal{K}(f)\colon \mathcal{K}(X,\mathcal{A}) \to \mathcal{K}(Y,\mathcal{B})$ is a homeomorphism between complexes satisfying strong conditions. A tico blowup corresponds to a certain kind of subdivision of $\mathcal{K}(X,\mathcal{A})$ or $\mathcal{K}(Y,\mathcal{B})$ by Theorem 7.4. You know when the map f lifts to a tico blowup by Theorem 7.5. So the problem reduces to finding subdivisions of $\mathcal{K}(X,\mathcal{A})$ and $\mathcal{K}(Y,\mathcal{B})$ so that $\mathcal{K}(f)$ sends vertices to vertices.

In the next section I give a proof along these lines in the dimension two case, a proof of the fact that a birational map can be factored into blowups and their inverses. I also give some partial results in dimension three. It seems to be a very interesting problem to determine whether this program can be worked out in general. Although Mori's program makes the dimension three problem not as useful as it was once thought, in higher dimensions it might be of some value. Moreover, the subdivision problem is eminently computable.

7. Proofs

In this section I give proofs or proof sketches of the results I talked about earlier. In this section I will assume some familiarity with notions defined in [**1**].

THEOREM 7.1. *The two definitions of* submersive *given in Definition* 2.1 *and Definition* 2.2 *are equivalent.*

PROOF. Suppose first that f is submersive, according to Definition 2.1 which is the definition used in [**1**]. Then an easy calculation shows that the logarithmic Jacobian has maximal rank everywhere. See the proof of Lemma 3.6.1 of [**1**].

Now suppose the logarithmic Jacobian has maximal rank everywhere. We will show that f is submersive according to Definition 2.1. Pick any point $p \in X$ and choose tico coordinates $\theta\colon (\mathbb{R}^m, 0) \to (X, p)$ and $\psi\colon (\mathbb{R}^n, 0) \to (Y, f(p))$ around p and $f(p)$. If f_i is the i-th coordinate of $\psi^{-1}f\theta$ then we have $f_i(x) = \prod_{j=1}^{\ell} x_j^{\beta_{ij}} \varphi_i(x)$ for all $i \leq k$ and $\varphi_i(0) \neq 0$. This is just what it means for f to be a tico map. Here the sheets of the tico in X near p are given by $x_i = 0$ for $i \leq \ell$ and the sheets of the tico in Y near $f(p)$ are given by $x_i = 0$ for $i \leq k$. The logarithmic Jacobian in these coordinates at the point p is:

$$\begin{pmatrix} \beta_{ij} & \partial \ln \varphi_i / \partial x_j \\ 0 & \partial f_i / \partial x_j \end{pmatrix}$$

This has rank n so the bottom right $(n-k) \times (m-\ell)$ submatrix has rank $n-k$. Consequently we may make a coordinate change in the last $m-\ell$ coordinates in X so that $f_i(x) = x_{i+m-n}$ for all $i > k$. This also shows that the stratum containing p, given locally by $x_1 = x_2 = \cdots = x_\ell = 0$, submerses to the stratum containing $f(p)$, given locally by $x_1 = x_2 = \cdots = x_k = 0$.

Let b be the rank of the upper left $k \times \ell$ matrix (β_{ij}). After reordering we may assume the first b rows of this matrix are linearly independent. Then after doing a coordinate change of the form $x_i' = x_i \prod_{j=1}^{b} \varphi_j(x)^{\gamma_{ij}}$ for $i \leq \ell$ and $x_i' = x_i$ for

$i > \ell$ we may assume that $\varphi_i(x) = 1$ for all $i \le b$. So the logarithmic Jacobian at p now looks like

$$\begin{pmatrix} \beta_{ij} & 0 & 0 \\ \beta_{ij} & \partial \ln \varphi_i/\partial x_j & \partial \ln \varphi_i/\partial x_j \\ 0 & 0 & I \end{pmatrix}$$

The middle $(k-b) \times (m-\ell-n+k)$ matrix must have rank $k-b$. After reordering the coordinates we may assume that the first $k-b$ columns of this middle matrix are linearly independent. Now do the coordinate change $x_i' = x_i$ if $i \le \ell$ or $i > \ell+k-b$ and $x_i' = \varphi_{i+b-\ell}$ if $\ell < i \le \ell+k-b$. After this coordinate change we get $\varphi_i(x) = x_{i-b+\ell}$ for all $b < i \le k$. Thus in these coordinates, f is a monomial map. Hence f is submersive according to the first definition. □

THEOREM 7.2. *The map resolution conjecture Theorem* 4.1 *is true if X has dimension 3.*

PROOF. An outline of my proof for the case $\dim X = 3$ is as follows. For each $k \le \dim Y$ there is a coherent sheaf $\mathcal{I}_k(f)$ of ideals of $\mathcal{O}(X)$ locally generated by the determinants of $k \times k$ minors of the logarithmic Jacobian of f. It is elementary to check that this ideal is independent of the coordinate chart chosen. Of course, f is submersive if and only if $\mathcal{I}_{\dim Y}(f) = \mathcal{O}(X)$.

The proof is in two stages. First find some sequence of blowups $\pi\colon X' \to X$ so that $\mathcal{I}_2(f \circ \pi)$ is locally principal. This is done using the Bierstone-Milman invariant of $\mathcal{I}_2$. Note it is quite easy, given a coherent sheaf of ideals, to use the Bierstone-Milman invariant to blow up and make it locally principal. This was the companion theorem to resolution of singularities in [**3**]. Its proof using the Bierstone-Milman invariant is essentially the same as their proof of the hypersurface case in [**2**] and thus is even easier than the proof of resolution of singularities since no special bases for ideals are needed. The only difference with the hypersurface case is that instead of starting out the invariant using $(g,1)$ for g a local equation of the hypersurface, you start it out with $\{(g_1,1),\dots,(g_k,1)\}$ where $g_1,\dots,g_k$ locally generate the ideal. Everything else is the same as the hypersurface case.

So at first glance, it would seem to be no trouble to do this first step of blowing up to make $\mathcal{I}_2$ locally principal. But it is not that easy. The problem is that $\mathcal{I}_2(f \circ \pi)$ is not exactly the transform $\pi^*\mathcal{I}_2(f)$ of the ideal $\mathcal{I}_2(f)$. It is close but not quite the same. But it is close enough that the Bierstone-Milman invariant can be used if $\dim X = 3$. Usually, this invariant will not increase after blowing up a center on which it is maximal, but sometimes it does increase. This only happens in very special situations which you can deal with on a case by case basis.

Anyway, after the first step you may as well assume that $\mathcal{I}_2(f)$ is locally principal. If $\dim Y = 3$ then $\mathcal{I}_3(f)$ is automatically locally principal also since there is only one 3×3 minor. So all $\mathcal{I}_k(f)$ are locally principal and their zeroes are ticos. Consequently, there are very particular local forms for the map f. One then uses some ad hoc invariants such as those used in the proof of Theorem 7.4.1 of [**1**] to blow up X and Y some more to make f submersive. □

THEOREM 7.3. *Let $f\colon (X,\mathcal{A}) \to (Y,\mathcal{B})$ be an exact tico map with constant exponents. Then there is an induced map $\mathcal{K}(f)\colon \mathcal{K}(X,\mathcal{A}) \to \mathcal{K}(Y,\mathcal{B})$ which satisfies the usual functorial properties $\mathcal{K}(f\circ g) = \mathcal{K}(f)\circ\mathcal{K}(g)$ and $\mathcal{K}(id) = id$.*

PROOF. If $A \in \mathcal{A}$ and $B \in \mathcal{B}$, let α_{AB} be the exponent of A in B for the map f. This means $f^*(\mathfrak{I}(B)) = \prod_{A\in\mathcal{A}} \mathfrak{I}(A)^{\alpha_{AB}}$. Let

$$\mathcal{B}_A = \{\, B\in\mathcal{B} \mid \alpha_{AB} > 0 \,\} = \{\, B\in\mathcal{B} \mid f(A)\subset B \,\}$$

Note $\mathcal{B}_A$ is nonempty if A is, since f is exact. Now $f(A)\subset \bigcap_{B\in\mathcal{B}_A} B$ so if $A\in\mathcal{A}$ is nonempty, then the vertices in $\mathcal{B}_A$ form a simplex σ in $\mathcal{K}(Y,\mathcal{B})$. Number the sheets in $\mathcal{B}_A$ is some way $B_1,\dots,B_\ell$. Let α_i be the exponent of A in B_i. Let $\mathcal{K}(f)(A)$ be the point in σ with projective barycentric coordinates $(\alpha_1 : \alpha_2 : \cdots : \alpha_\ell)$.

It might seem natural at this point to extend linearly to obtain a PL map. However, if one did so the composition rule would not hold. What we do instead is the following. Take a point p in a simplex δ of $\mathcal{K}(X,\mathcal{A})$. Denote the vertices of δ as $A_1,\dots,A_k$ and suppose p has barycentric coordinates $(t_0,\dots,t_k)$. Let

$$\mathcal{B}_\delta = \bigcup_{i=1}^{k} \mathcal{B}_{A_i} = \{\, B\in\mathcal{B} \mid f(\bigcap_{i=1}^{k} A_i)\subset B \,\}$$

Let σ be the simplex of $\mathcal{K}(Y,\mathcal{B})$ with vertices $\mathcal{B}_\delta$. Number the sheets of $\mathcal{B}_\delta$ in some way $B_1,\dots,B_\ell$. Let α_{ji} be the exponent of A_j in B_i. Then $\mathcal{K}(f)(p)$ is the point of σ with projective barycentric coordinates $(\beta_1 : \cdots : \beta_\ell)$ where $\beta_i = \sum_{j=1}^{k} t_j\alpha_{ji}$. It is elementary to check that the functorial properties are satisfied.

Note that the image of a simplex is the convex hull of the images of its vertices. So on that level it looks like a PL map. It is just that the map $\mathcal{K}(f)$ itself is not PL, because radial projection is not a PL map. □

There is a more concrete construction of $\mathcal{K}(X,\mathcal{A})$ and $\mathcal{K}(f)$ which the reader may find helpful. We assume $\mathcal{A}$ is finite. Suppose there are n nonempty sheets in $\mathcal{A}$. Then we can realize $\mathcal{K}(X,\mathcal{A})$ as a subset of $\mathbb{R}^n$ as follows. Order the nonempty sheets of $\mathcal{A}$ as $A_1,\dots,A_n$. Let e_i be the standard basis vector in $\mathbb{R}^n$. Then $\{\, x\in\mathbb{R}^n \mid x_i \geq 0 \text{ and } \sum_{i=1}^{n} x_i = 1 \,\}$ is a standard $(n-1)$-simplex, the convex hull of $e_1,\dots,e_n$. Then $\mathcal{K}(X,\mathcal{A})$ is the union of the of the faces of this simplex corresponding to nonempty intersections of the A_i's. In other words, $|\mathcal{K}(X,\mathcal{A})|$ is the union of convex hulls of $e_{i_1},\dots,e_{i_k}$ where $\bigcap_{j=1}^{k} A_{i_j}$ is nonempty. If $\mathcal{B}$ has m nonempty sheets we have an $m\times n$ matrix L of exponents of the map f. Then $\mathcal{K}(f)$ is the composition of L and radial projection to the affine hyperplane $\sum_{i=1}^{m} y_i = 1$.

Let us now see the effect of a tico blowup on this simplicial complex. It is not hard to see the effect in general, but we will restrict our attention to the case where the center of the blowup is an intersection of sheets. In this case, the effect is to subdivide $\mathcal{K}(X, \mathcal{A})$ in a certain way. In particular:

THEOREM 7.4. *Let $\pi: (X', \mathcal{A}') \to (X, \mathcal{A})$ be a tico blowup with nonempty center $C = \bigcap_{i=0}^{k} A_i$ for some sheets $A_i \in \mathcal{A}$. Then $\mathcal{K}(X', \mathcal{A}')$ is isomorphic to a subdivision of the simplicial complex $\mathcal{K}(X, \mathcal{A})$. This subdivision is as follows. Let σ be the k-simplex of $\mathcal{K}(X, \mathcal{A})$ with vertices $A_0, \dots, A_k$. Then $\mathcal{K}(X', \mathcal{A}')$ is obtained from $\mathcal{K}(X, \mathcal{A})$ by replacing the closed star of σ with the cone on its boundary.*

PROOF. We may suppose that $k > 0$, otherwise $(X, \mathcal{A})$ and $(X', \mathcal{A}')$ are isomorphic and the result is trivial. Consider the map $\mathcal{K}(\pi): |\mathcal{K}(X', \mathcal{A}')| \to |\mathcal{K}(X, \mathcal{A})|$. The exponents of the tico map π are given in Lemma 3.3.4 of [**1**]. If $A \in \mathcal{A}$ is a nonempty sheet, then $\mathcal{K}(\pi)$ sends the vertex the strict transform of A to the vertex A. Also, $\mathcal{K}(\pi)$ sends the sheet $\pi^{-1}(C)$ to the barycenter of σ. It is elementary to check that $\mathcal{K}(\pi)$ is a homeomorphism and performs the indicated subdivision. □

We have the following criterion for lifting maps.

THEOREM 7.5. *Let $f: (X, \mathcal{A}) \to (Y, \mathcal{B})$ be an exact tico map with constant exponents and let $\pi: (Y', \mathcal{B}') \to (Y, \mathcal{B})$ be a tico blowup whose center C is an intersection of sheets. Then we may lift f to a tico map $g: (X, \mathcal{A}) \to (Y', \mathcal{B}')$ with $f = \pi \circ g$ if and only if for each simplex σ of $\mathcal{K}(X, \mathcal{A})$, the image $\mathcal{K}(\pi)^{-1} \circ \mathcal{K}(f)(\sigma)$ is contained in a simplex of $\mathcal{K}(Y', \mathcal{B}')$.*

PROOF. If g exists, then $\mathcal{K}(\pi)^{-1} \circ \mathcal{K}(f) = \mathcal{K}(g)$ which takes simplices into simplices. So this implication is trivial.

Now suppose that $\mathcal{K}(\pi)^{-1} \circ \mathcal{K}(f)$ takes simplices into simplices. Let $B_0, \dots, B_k$ be the sheets of $\mathcal{B}$ so that $C = \bigcap_{i=0}^{k} B_i$. Take a point $q \in X$ so that there is no local lifting of f near q. Let $A_0, \dots, A_\ell$ be the sheets of $\mathcal{A}$ which contain q. Let σ be the ℓ-simplex of $\mathcal{K}(X, \mathcal{A})$ with vertices $A_0, \dots, A_\ell$. Let δ be the smallest simplex of $\mathcal{K}(Y, \mathcal{B})$ which contains $\mathcal{K}(f)(\sigma)$. Let δ have dimension m. Note that we must have $f(q) \in C$ since f does not lift near q, so for each B_i we must have the exponent of A_j in B_i is greater than zero for some j. Consequently, each B_i must be a vertex of δ. We may thus name the vertices of δ as $B_0, \dots, B_m$. The map $\mathcal{K}(\pi)$ subdivides δ as indicated in Theorem 7.4. There are $k+1$ new m-simplices in the subdivision, the i-th one consists of all points of δ whose i-th barycentric coordinate is less than or equal to all the j-th barycentric coordinates for $0 \leq j \leq k$. Consequently, after reordering we may assume that for each point in $\mathcal{K}(f)(\sigma)$, its 0-th barycentric coordinate is less than or equal to all the j-th barycentric coordinates for $0 \leq j \leq k$. In particular, looking at $\mathcal{K}(f)(A_i)$ we see that if α_{ij} is the exponent of A_i in B_j, then $\alpha_{i0} \leq \alpha_{ij}$ for all $0 \leq j \leq k$. Now $\mathfrak{I}(C) = \sum_{i=0}^{k} \mathfrak{I}(B_i)$ so locally we have

$$f^*(\mathfrak{I}(C)) = \sum_{j=0}^{k} \prod_{i=0}^{\ell} \mathfrak{I}(A_i)^{\alpha_{ij}} = \prod_{i=0}^{\ell} \mathfrak{I}(A_i)^{\alpha_{i0}}$$

so $f^*(\mathfrak{I}(C))$ is locally principal and f lifts near q, a contradiction. □

LEMMA 7.1. *Let $f\colon (X, \mathcal{A}) \to (Y, \mathcal{B})$ be an exact tico map with constant exponents. Suppose the map $\mathcal{K}(f)$ takes the interior of a simplex δ into the interior of a simplex σ. (Thus σ is the smallest simplex of $\mathcal{K}(Y, \mathcal{B})$ which contains $\mathcal{K}(f)(\delta)$.) Suppose δ has vertices $A_1, \ldots, A_k$ and σ has vertices $B_1, \ldots, B_\ell$. Pick some point $p \in \bigcap_{i=1}^{k} A_i$ which is in no other sheets of $\mathcal{A}$. Then $f(p)$ is in the sheets B_i and in no other sheets of $\mathcal{B}$.*

PROOF. Suppose $f(p) \in B$, for some sheet B in $\mathcal{B}$. We know that $f^{-1}(B)$ is a union of sheets by Lemma 3.2.1 of [**1**] Since $p \in f^{-1}(B)$ and p is only in the A_i's we know that one of the sheets A_i is in $f^{-1}(B)$. But then Lemma 3.2.2 of [**1**] says A_i must have a positive exponent with B, so $\mathcal{K}(f)(A_i)$ has nontrivial B coordinate and B must be a vertex of σ. □

LEMMA 7.2. *Let $f\colon (X, \mathcal{A}) \to (Y, \mathcal{B})$ be a submersive exact tico map with constant exponents between complex ticos and suppose that f is a proper analytic map. Suppose the map $\mathcal{K}(f)$ takes the interior of a simplex δ into the interior of a simplex σ. Let $A_0, \ldots, A_n$ be the vertices of δ and $B_0, \ldots, B_m$ be the vertices of σ. Then $f(\bigcap_{i=0}^{n} A_i)$ is a union of connected components of $\bigcap_{i=0}^{m} B_i$. In the algebraic case, $f(\bigcap_{i=0}^{n} A_i)$ is a union of irreducible components of $\bigcap_{i=0}^{m} B_i$.*

PROOF. Since f is proper we know that $f(\bigcap_{i=0}^{n} A_i)$ is a subvariety of Y. Since $\mathcal{K}(f)$ takes the interior of δ to the interior of σ we know that for each $i \leq m$ there is a $j \leq n$ so that the exponent of A_j in B_i is nonzero. Consequently, $f(A_j) \subset B_i$. But then $f(\bigcap_{i=0}^{n} A_i) \subset \bigcap_{i=0}^{m} B_i$. So we only need show that $f(\bigcap_{i=0}^{n} A_i)$ is an open subset of $\bigcap_{i=0}^{m} B_i$. This follows from properness and the fact that f submerses strata to strata. □

Now let us see what happens to a submersive tico map with constant exponents which is also a proper birational equivalence.

THEOREM 7.6. *Let $f\colon (X, \mathcal{A}) \to (Y, \mathcal{B})$ be an exact tico map with constant exponents between complex ticos and suppose that f is proper and a birational equivalence. Suppose also that Y is irreducible and any intersection of sheets of $\mathcal{B}$ is either empty or irreducible. Then the map $\mathcal{K}(f)$ is onto. If, in addition, f is submersive then $\mathcal{K}(f)$ is a homeomorphism.*

PROOF. Pick an algebraic subvariety $Z \subset Y$ and a rational function $g\colon Y - Z \to X$ so that $f \circ g$ is the identity on $Y - Z$.

First we see that $\mathcal{K}(f)$ is onto. Take any simplex of $\mathcal{K}(Y,\mathcal{B})$. Denote its vertices $B_1,\ldots,B_k$. Take some point c in the interior of this simplex with rational barycentric coordinates. Let its barycentric coordinates coordinates be $(c_1/d,\ldots,c_k/d)$ for some positive integers c_i and d.

Take some point $p \in \bigcap_{i=1}^k B_i$ so that p is in no sheet of $\mathcal{B}$ except for the B_i's. Take analytic tico coordinates $\theta\colon (U,0) \to (Y,p)$ around p where U is an open subset of $\mathbb{C}^n$ so that $\theta^{-1}(B_i) = \mathbb{C}_i^n \cap U$. For each $z \in \mathbb{C}^n$ consider the analytic curve

$$\alpha_z(t) = \theta(t^{c_1} + z_1 t^{2d}, t^{c_2} + z_2 t^{2d}, \ldots, t^{c_k} + z_k t^{2d}, z_{k+1}t^{2d}, \ldots, z_n t^{2d})$$

Note that the union of the images of these curves is open and hence for some z near 0 and $\epsilon > 0$, we have $\alpha_z(t) \notin Z$ for all $t \in \mathbb{C}$ with $0 < |t| < \epsilon$. Then by properness of f, there is a limit point q of $g\alpha_z(t)$ as $t \to 0$. But then $q \cup g\alpha_z(\epsilon\mathbb{C} - 0)$ is an analytic curve in X. (It is an irreducible component at q of the analytic set $f^{-1}(\alpha_z(\epsilon\mathbb{C})) = g\alpha_z(\epsilon\mathbb{C} - 0) \cup f^{-1}(p)$.)

Denote by $A_1,\ldots,A_\ell$ the sheets of $\mathcal{A}$ which contain q. Pick any analytic tico coordinates $\psi\colon (V,0) \to (X,q)$ in a neighborhood of q where V is open in $\mathbb{C}^n$ so that $\psi^{-1}(A_i) = \mathbb{C}_i^n$ for all $i \le \ell$. Since f is a tico map, we know that if f_i is the i-th coordinate of $\theta^{-1} f \psi$ then $f_i(x) = \prod_{j=1}^{\ell} x_j^{\beta_{ij}} \varphi_i(x)$ for all $i \le k$. Here β_{ij} are some nonnegative integers and $\varphi_i(0) \ne 0$.

Take an analytic parameterization of the analytic curve $\psi^{-1}(q \cup g\alpha_z(\epsilon\mathbb{C} - 0))$, say $\gamma(s) = (\gamma_1(s),\ldots,\gamma_m(s))$ for analytic γ_i's. Since α_z does not intersect $|\mathcal{B}|$ near p except at p and since $f^{-1}(|\mathcal{B}|) = |\mathcal{A}|$, we know for each $i \le \ell$ that γ_i is not identically 0. So let e_i be the order of γ_i for $i \le \ell$ and let $\gamma_i'(s) = s^{-e_i}\gamma_i(s)$. Note $\gamma_j'(0) \ne 0$. Now $\theta^{-1} f \psi \gamma$ is another parameterization of the curve $\theta^{-1}\alpha_z$. For $i \le k$, the i-th coordinate of $\theta^{-1} f \psi \gamma(s)$ is

$$\left(s^{\sum_{j=1}^{\ell} \beta_{ij} e_j}\right) \prod_{j=1}^{\ell} \gamma_j'(s)^{\beta_{ij}} \varphi_i(\gamma(s))$$

Consequently, comparing the relative orders of the first k coordinates in the two parameterizations, we see that there is a rational number r so that for each $i \le k$ we have $\sum_{j=1}^{\ell} \beta_{ij} e_j = r c_i$. But this means exactly that if we take the simplex in $\mathcal{K}(X,\mathcal{A})$ with vertices $A_1,\ldots,A_\ell$ and take the point in it with projective barycentric coordinates $(e_1 : \ldots : e_\ell)$ and apply $\mathcal{K}(f)$, then we get the point c. Consequently, $\mathcal{K}(f)$ is onto since the points with rational barycentric coordinates are dense in $\mathcal{K}(Y,\mathcal{B})$.

Now we will show that $\mathcal{K}(f)$ is one to one if f is submersive. Suppose the images of two different simplices intersect. Then it is easy to see that they must intersect in some point c which has rational barycentric coordinates. This point c then has (at least) two different inverse images e and e' which also have rational barycentric coordinates. Let e be in the interior of a simplex with vertices $A_1,\ldots,A_\ell$ and let e' be in the interior of a simplex with vertices $A_1',\ldots,A_{\ell'}'$. Let c be in the interior of a simplex with vertices $B_1,\ldots,B_k$.

Pick any point $p \in \bigcap_{i=1}^k B_i$ so that p is in no other sheets of $\mathcal{B}$. By Lemma 7.2, we know that $f(\bigcap_{i=1}^{\ell} A_i) = \bigcap_{i=1}^k B_i$ and $f(\bigcap_{i=1}^{\ell'} A'_i) = \bigcap_{i=1}^k B_i$. So we may pick $q \in \bigcap_{i=1}^{\ell} A_i$ and $q' \in \bigcap_{i=1}^{\ell'} A'_i$ so that $f(q) = f(q') = p$.

Pick a curve α_z as we did above. I claim we may pick curves β and β' near q and q' so that $f \circ \beta = f \circ \beta' = \alpha_z$. This then is a contradiction since both curves β and β' must be $g\alpha_z$, but they are distinct (even if $q = q'$).

To see the claim, I will construct β, the curve β' is constructed similarly. By the complex version of Lemma 3.6.2 of [**1**] we may find analytic tico coordinates $\psi\colon V \to X$ in a neighborhood of q where V is open in $\mathbb{C}^n$ so that $\psi^{-1}(A_i) = \mathbb{C}^n_i$ for all $i \leq \ell$ and if f_i is the i-th coordinate of $\theta^{-1} f \psi$ then $f_i(x) = x_i$ for all $i > k$ and $f_i(x) = \prod_{j=1}^k x_j^{\beta_{ij}}$ for all $i \leq k$. Here β_{ij} are some nonnegative integers. Note that the i-th coordinate of $\psi^{-1}(q)$ is 0 for $i \leq \ell$ and $i > k$ but is nonzero for all other i's. Choose positive integers e_i so that e has projective barycentric coordinates $(e_1 : \ldots : e_\ell)$. Our curve β will be of the form

$$\beta(s) = \psi(s^{e_1}\gamma_1(s), \ldots, s^{e_\ell}\gamma_\ell(s), \gamma_{\ell+1}(s), \ldots, \gamma_n(s))$$

where $\gamma_i(0) \neq 0$ for all $i \leq k$. Since $\mathcal{K}(f)(e) = c$ we know there is a rational number r so that $\sum_{j=1}^{\ell} \beta_{ij} e_j = rc_i$ for each $i \leq k$. In fact, for convenience, choose the e_i so that r is an integer. So we want to solve the equation $\theta^{-1} f \beta(s) = \theta^{-1}\alpha_z(s^r)$. This becomes the equations

$$\begin{array}{rcll} \prod_{j=1}^k \gamma_j(s)^{\beta_{ij}} & = & 1 + z_i s^{2rd - rc_i} & \text{for } i \leq k \\ \gamma_i(s) & = & z_i s^{2rd} & \text{for } i > k \end{array}$$

Since the matrix (β_{ij}) is onto there are rational numbers κ_{ja} so that $\sum_{j=1}^k \beta_{ij}\kappa_{ja}$ is the Kronecker delta δ_{ia}. We may then let

$$\gamma_j(s) = \begin{cases} \prod_{i=1}^k (1 + z_i s^{2rd - rc_i})^{\kappa_{ji}} & \text{for } j \leq k \\ z_j s^{2rd} & \text{for } j > k \end{cases}$$

and we have found our curve β. □

THEOREM 7.7. *Let $f\colon (X, \mathcal{A}) \to (Y, \mathcal{B})$ be a submersive exact tico map with constant exponents between complex ticos and suppose that f is proper and a birational equivalence. Let σ be an $\ell - 1$ simplex in $\mathcal{K}(X, \mathcal{A})$ and let δ be the smallest simplex of $\mathcal{K}(Y, \mathcal{B})$ which contains $\mathcal{K}(f)(\sigma)$. Let δ have dimension $k-1$. We have a $k \times \ell$ matrix T formed from the exponents of the vertices of σ in the vertices of δ. (So the columns of T give projective barycentric coordinates for the images of the vertices of σ.)*

Then $T^{-1}(\mathbb{Z}^k) \subset \mathbb{Z}^\ell$, the inverse image of the integer lattice is contained in the integer lattice. Moreover, T is a submatrix of a $k \times k$ nonnegative integer matrix T' with determinant ± 1. Consequently each column of T has greatest common divisor 1.

PROOF. Let the vertices of σ be $A_1, \dots, A_\ell$ and let $B_1, \dots, B_k$ be the vertices δ. Pick some point $q \in \bigcap_{i=1}^k A_i$ which is in no other sheets of $\mathcal{A}$. Then by Theorem 7.1, $f(q)$ is in the sheets B_i and in no other sheets of $\mathcal{B}$.

Take analytic tico coordinates $\theta\colon (U,0) \to (Y, f(q))$ around $f(q)$ where U is an open subset of $\mathbb{C}^n$ so that $\theta^{-1}(B_i) = \mathbb{C}^n_i \cap U$. By the complex version of Lemma 3.6.2 of [**1**] we may find analytic tico coordinates $\psi\colon V \to X$ in a neighborhood of q where V is open in $\mathbb{C}^n$ so that $\psi^{-1}(A_i) = \mathbb{C}^n_i$ for all $i \leq \ell$ and if f_i is the i-th coordinate of $\theta^{-1} f \psi$ then $f_i(x) = x_i$ for all $i > k$ and $f_i(x) = \prod_{j=1}^k x_j^{\beta_{ij}}$ for all $i \leq k$. Here β_{ij} are some nonnegative integers. Note that the i-th coordinate of $\psi^{-1}(q)$ is 0 for $i \leq \ell$ and $i > k$ but is nonzero for all other i's.

Note that the ij-th entry of T is β_{ij}. Suppose that $Tc \in \mathbb{Z}^k$ for some $c \notin \mathbb{Z}^\ell$. Then for any x near $\psi^{-1}(q)$ in V, we have $f\psi(x) = f\psi(x')$ where $x'_i = x_i$ for $i > \ell$ and $x'_i = e^{2\pi c_i \sqrt{-1}} x_i$ for $i \leq \ell$. This is a contradiction to birationality of f, since f must be locally generically one to one. So we know that $T^{-1}(\mathbb{Z}^k) \subset \mathbb{Z}^\ell$.

If $\ell = k$ then this implies T has determinant ± 1. Since $\mathcal{K}(f)$ is a homeomorphism, we know that σ is a face of a $k-1$ simplex σ' so that $\mathcal{K}(f)(\sigma') \subset \delta$. The matrix of exponents T' we associate to σ' has determinant ± 1 and its first ℓ columns are exactly T. □

THEOREM 7.8. *Let $f\colon (X, \mathcal{A}) \to (Y, \mathcal{B})$ be a submersive tico map satisfying the conditions of Theorem* 7.6. *Suppose also that $\mathcal{K}(f)$ takes vertices to vertices. Then f has an entire rational inverse.*

PROOF. Let the sheets of $\mathcal{A}$ be $A_1, \dots A_k$. Then we may write the sheets as $\mathcal{B}$ as $B_1, \dots B_k$ where $\mathcal{K}(f)(A_i) = B_i$. By Lemma 7.2 we know that $f(A_i) = B_i$. By Lemma 3.2.2 of [**1**] we also know $A_i = f^{-1}(B_i)$. Since the exponent of A_i in B_j is the Kronecker delta, the local form of the submersive tico map is by definition the identity, so f is locally an analytic isomorphism. But then since it is birational, it must be globally an isomorphism. □

Although Lemma 7.3 and Lemma 7.4 below ask for the hypotheses of Theorem 7.6, in reality they require less, namely that $\mathcal{K}(f)$ be a homeomorphism and that the conclusion of Theorem 7.7 hold, more particularly, Lemma 7.3 requires that if $\mathcal{K}(f)$ maps a 1-simplex into a 1-simplex then the 2×2 matrix of exponents have determinant ± 1 and Lemma 7.4 requires that if $\mathcal{K}(f)$ maps a 2-simplex into a 2-simplex then the 3×3 matrix of exponents have determinant ± 1.

This brings to mind the following question. Suppose $(Y, \mathcal{B})$ is such that all intersections of sheets are irreducible and suppose we have a subdivision of $\mathcal{K}(Y, \mathcal{B})$ satisfying the conclusions of Theorem 7.7, i.e., we have choices of integer projective barycentric coordinates of all vertices so that all $k \times k$ matrices formed from vertices of $k-1$ simplices in a $k-1$ simplex of $\mathcal{K}(Y, \mathcal{B})$ will have determinant ± 1. Then is there an $(X, \mathcal{A})$ and a birational submersive tico map $f\colon (X, \mathcal{A}) \to (Y, \mathcal{B})$ giving this subdivision? This is true in dimensions less than or equal to 2.

LEMMA 7.3. *Let $f\colon (X,\mathcal{A}) \to (Y,\mathcal{B})$ be a tico map satisfying the conditions of Theorem 7.6. Let σ be a 1-simplex of $\mathcal{K}(Y,\mathcal{B})$ and suppose that $\mathcal{K}(f)$ takes some vertex to the interior of σ. Then $\mathcal{K}(f)$ takes some vertex to the barycenter of σ.*

PROOF. Let σ have vertices B_1 and B_2. Let c denote the barycenter of σ. Blow up the point $B_1 \cap B_2$ in Y obtaining a tico blowup $\pi\colon (Y',\mathcal{B}') \to (Y,\mathcal{B})$. By Proposition 3.3.9 of **[1]** we know that after some sequence of tico blowups $\rho\colon (X,\mathcal{A}') \to (X,\mathcal{A})$ we have a lifting $g\colon X' \to Y'$ so $\pi g = f\rho$. Moreover, the proof of Proposition 3.3.9 shows that the centers of all blowups are intersections of two sheets. We will now show that this sequence of blowups is not necessary, i.e., we could take $\rho =$ identity.

Take a ρ which has a smallest number of blowups. Suppose that some blowing up is necessary. Consider the last blowup in ρ. So $\rho = \rho'' \circ \rho'$ where $\rho'\colon (X',\mathcal{A}') \to (X'',\mathcal{A}'')$ and $\rho''\colon (X'',\mathcal{A}'') \to (X,\mathcal{A})$ and ρ' blows up the intersection of two sheets A_1 and A_2 of $\mathcal{A}''$. Let $d \in |\mathcal{K}(X'',\mathcal{A}'')|$ be the barycenter of the simplex with vertices A_1 and A_2. By minimality of ρ we know that $f \circ \rho''$ does not lift, but $f \circ \rho$ does lift. Hence by Theorem 7.4 and Theorem 7.5 we know that $\mathcal{K}(f\circ\rho'')(d) = c$. Let $\mathcal{K}(f\circ\rho'')(A_i)$ have projective barycentric coordinates $(\beta_{1i} : \beta_{2i})$, in other words β_{ji} are the exponents of A_i in B_j for $f\circ\rho''$. By Theorem 7.7, we know that $\beta_{11}\beta_{22} - \beta_{12}\beta_{21} = \pm 1$. On the other hand, $\mathcal{K}(f \circ \rho'')(d) = c$. But $\mathcal{K}(f \circ \rho'')(d)$ has projective barycentric coordinates $(\beta_{11} + \beta_{12} : \beta_{21} + \beta_{22})$ so $\beta_{11} + \beta_{12} = \beta_{21} + \beta_{22}$. But then by Theorem 7.7, we have $\beta_{11} + \beta_{12} = \beta_{21} + \beta_{22} = 1$. So the only two possibilities for the integers β_{ji} are $\beta_{11} = \beta_{22} = 1$ and $\beta_{12} = \beta_{21} = 0$ or $\beta_{11} = \beta_{22} = 0$ and $\beta_{12} = \beta_{21} = 1$. In other words, the vertices A_i get mapped to the vertices B_j. This together with the fact that $\mathcal{K}(f\circ\rho'')$ is a homeomorphism implies that no vertex of $\mathcal{K}(X'',\mathcal{A}'')$ maps to the interior of the simplex B_1B_2. In particular, no vertex of $\mathcal{K}(X,\mathcal{A})$ is mapped to the interior so we have a contradiction. Consequently, the minimal ρ is just the identity which is what we wanted.

So the map f lifts. Take a simplex δ so that $c \in \mathcal{K}(f)(\delta)$. Then Theorem 7.4 and Theorem 7.5 imply that the image of δ lies on one side or the other of c. In particular, some vertex of δ must map to c. □

Let us use these facts to show that a proper, birational map between complex surfaces is a composition of blowings up and blowings down. So let $f\colon X \to Y$ be a proper birational map between complex surfaces. We may as well assume that Y is irreducible. By Theorem 4.1, after blowing up X and Y we may as well assume there are algebraic ticos $\mathcal{A}$ in X and $\mathcal{B}$ in Y so that f is a submersive tico map. After blowing up irreducible components of intersections of sheets in Y, we may also assume that intersections of sheets in $\mathcal{A}$ and $\mathcal{B}$ are all irreducible. Lemma 3.6.3 of **[1]** shows that this extra blowing up does not destroy submersiveness. Then we may apply Theorem 7.6 to see that $\mathcal{K}(f)$ gives a homeomorphism from $|\mathcal{K}(X,\mathcal{A})|$ to $|\mathcal{K}(Y,\mathcal{B})|$.

Since X and Y are surfaces, all threefold intersections of sheets are empty, so $\mathcal{K}(X,\mathcal{A})$ and $\mathcal{K}(Y,\mathcal{B})$ both have dimensions ≤ 1. Suppose $\mathcal{K}(f)$ maps some vertex of $\mathcal{K}(X,\mathcal{A})$ to the interior of a 1-simplex of $\mathcal{K}(Y,\mathcal{B})$. Let this 1-simplex have vertices B_1 and B_2. By Lemma 7.3, some vertex of $\mathcal{K}(X,\mathcal{A})$ is mapped to the barycenter of σ. Hence every simplex which maps to σ must map to one side or the other of the barycenter. Blow up the point $B_1 \cap B_2$ in Y obtaining a tico blowup $\pi\colon (Y',\mathcal{B}') \to (Y,\mathcal{B})$. By Theorem 7.5, we have a lifting $g\colon X \to Y'$ so $\pi g = f$. We continue to blow up Y as long as there are vertices of $\mathcal{K}(X,\mathcal{A})$ which are not mapped to vertices. Eventually this process must stop since after each step there is one more vertex of $\mathcal{K}(X,\mathcal{A})$ which is mapped to a vertex. At the end we have a lifting $g'\colon (X,\mathcal{A}) \to (Y'',\mathcal{B}'')$ and $\mathcal{K}(g')$ takes vertices to vertices. By Lemma 3.6.3 of [**1**] and the following Theorem 7.8 then g' is an algebraic isomorphism. In other words, a proper submersive exact f is in fact a blowup map, if intersections of sheets are irreducible.

The next question is what happens in dimension 3. We can try the same idea there. Again we may assume that f is submersive. However, it turns out that sometimes we need to do the blowups ρ of X so a proof along these lines must be more complicated. However, the above argument takes care of the one dimensional parts of $\mathcal{K}(X,\mathcal{A})$ and we can direct our full attention to the 2-simplices. We then have the following result.

LEMMA 7.4. *Let $f\colon (X,\mathcal{A}) \to (Y,\mathcal{B})$ be a tico map satisfying the conditions of Theorem* 7.6. *Let σ be a 2-simplex of $\mathcal{K}(Y,\mathcal{B})$ with vertices B_1, B_2, B_3. Let A_1, A_2 and A_3 be the unique sheets of $\mathcal{A}$ so that $\mathcal{K}(f)(A_i) = B_i$, $i = 1,2,3$. Then exactly one of the following possibilities holds.*

(i) *Some 2-simplex of $\mathcal{K}(X,\mathcal{A})$ maps onto σ. Note that this simplex must be $A_1A_2A_3$.*
(ii) *There is a sheet A_4 of $\mathcal{A}$ so that $\mathcal{K}(f)(A_4)$ is the barycenter of σ.*
(iii) *After perhaps reordering the B_i's, there is a sheet A_5 in $\mathcal{A}$ so that $\mathcal{K}(f)(A_5)$ is the barycenter of B_2 and B_3 and so A_1 and A_5 form a simplex in $\mathcal{K}(X,\mathcal{A})$.*

PROOF. Suppose that neither the first nor the second possibility happens. First, we may assume that some sheet in $\mathcal{K}(X,\mathcal{A})$ maps to the interior of one of the edges of σ. For suppose it does not. Then A_2A_3 is a simplex of $\mathcal{K}(X,\mathcal{A})$ and so $A_3 \cap A_2$ is nonempty. Let $\rho\colon (X',\mathcal{A}') \to (X,\mathcal{A})$ be the blow up with center $A_3 \cap A_2$. After the blowup, we still have no sheets mapping to the barycenter of σ, but we do have a sheet mapping to the interior of B_3B_2 so by the following argument, the last possibility (iii) holds. So A_1', the strict transform of A_1, and $A_5 = \rho^{-1}(A_2 \cap A_3)$ form a simplex in $\mathcal{K}(X',\mathcal{A}')$. But this means that $A_1 \cap A_2 \cap A_3$ is nonempty which is a contradiction since then there would be a simplex $A_1A_2A_3$ which $\mathcal{K}(f)$ would map onto σ and the first possibility (i) would hold.

So we may assume, after reordering, that $\mathcal{K}(f)$ maps some sheet to the interior of the edge B_2B_3. By Lemma 7.3, there is a sheet A_6 of $\mathcal{A}$ so that $\mathcal{K}(f)(A_6)$ is the barycenter of B_2B_3. Let $\pi\colon (Y',\mathcal{B}') \to (Y,\mathcal{B})$ be the tico blowup with center $B_2 \cap B_3$. By Proposition 3.3.9 of [1], there is a sequence of tico blowups $\rho\colon (X',\mathcal{A}') \to (X,\mathcal{A})$ and a tico map $f'\colon (X',\mathcal{A}') \to (Y',\mathcal{B}')$ so that $\pi f' = f\rho$. Also each center of each blowup in ρ is the intersection of two sheets. Now there is a 1-simplex σ'' in $\mathcal{K}(Y',\mathcal{B}')$ which $\mathcal{K}(\pi)$ maps to the line from B_1 to the barycenter of the edge B_2B_3. Note $\mathcal{K}(\pi)$ takes the barycenter of σ'' to the barycenter of σ.

Suppose that some vertex of $\mathcal{K}(X',\mathcal{A}')$ is mapped by $\mathcal{K}(f')$ to the interior of σ''. Then by Lemma 7.3, there is a sheet $A \in \mathcal{A}'$ so that $\mathcal{K}(f')(A)$ is the barycenter of σ''. Then $\mathcal{K}(f)(\mathcal{K}(\rho)(A))$ is the barycenter of σ so $\mathcal{K}(\rho)(A)$ cannot be a vertex of $\mathcal{K}(X,\mathcal{A})$. So we may factor ρ as the composition $\rho = \rho'' \circ \rho' \circ \rho'''$ where $\rho'''\colon (X',\mathcal{A}') \to (X''',\mathcal{A}''')$ and $\rho'\colon (X''',\mathcal{A}''') \to (X'',\mathcal{A}'')$ and $\rho''\colon (X'',\mathcal{A}'') \to (X,\mathcal{A})$ and ρ' is a single blowup with center C and $\mathcal{K}(\rho''')(A)$ is the vertex $A' = \rho'^{-1}(C)$.

Let $C = A_1'' \cap A_2''$ for sheets $A_i'' \in \mathcal{A}''$. Let β_{ji} be the exponent of A_i'' in B_j for $f \circ \rho''$. So $\mathcal{K}(f \circ \rho'')(A_i'')$ has projective barycentric coordinates $(\beta_{1i} : \beta_{2i} : \beta_{3i})$. Also $\beta_{j1} + \beta_{j2}$ is the exponent of A' in B_j for the map $f \circ \rho'' \circ \rho'$. Since $\mathcal{K}(f \circ \rho'' \circ \rho')(A')$ is the barycenter of σ, we must have $\beta_{11} + \beta_{12} = \beta_{21} + \beta_{22} = \beta_{31} + \beta_{32}$. Hence by Theorem 7.1, we must have $\beta_{11} + \beta_{12} = \beta_{21} + \beta_{22} = \beta_{31} + \beta_{32} = 1$. Consequently we know, after perhaps exchanging A_1'' and A_2'', that $\mathcal{K}(f \circ \rho'')(A_1'')$ is one of the vertices of σ and $\mathcal{K}(f \circ \rho'')(A_2'')$ is the barycenter of the opposite side. Hence after reordering we may assume that $\mathcal{K}(\rho'')(A_1'') = A_1$ and $\mathcal{K}(f \circ \rho'')(A_2'')$ is the barycenter of B_2B_3.

If $\mathcal{K}(\rho'')(A_2'')$ is a vertex, let this vertex be the sheet A_5. Since $A_1'' \cap A_2''$ is nonempty, we know that $A_1 \cap A_5$ is nonempty also so we are finished. So we are left with the case where $\mathcal{K}(\rho'')(A_2'')$ is not a vertex. Then Lemma 7.3 implies that $A_2 \cap A_3$ is nonempty and $\mathcal{K}(\rho'')(A_2'')$ is the barycenter of the 1-simplex A_2A_3. So at some point in the sequence of blowups ρ'', we blow up the intersection of the strict transforms of A_2 and A_3, and then A_2'' is the strict transform of the resulting exceptional divisor. But then, we would have $A_1 \cap A_2 \cap A_3$ nonempty which is a contradiction. We could also see this geometrically, since $\mathcal{K}(\rho'')$ gives a subdivision of the complex $\mathcal{K}(X,\mathcal{A})$ and this subdivision includes a one simplex from A_1 to the barycenter of A_2A_3. Hence $A_1A_2A_3$ must have been a simplex which is a contradiction.

One can give another proof which the reader may prefer. Pick a 2-simplex δ so that $\mathcal{K}(f)(\delta)$ contains the barycenter of $B_1B_2B_3$. Let δ have vertices $A_6A_7A_8$. Let T_i be the 3×1 vector whose entries are the exponents of A_i in B_1, B_2 and B_3. If T is the 3×3 matrix whose columns are T_6, T_7 and T_8 then we know that T is invertible over the integers so there are unique integers a_i so that $a_6T_6 + a_7T_7 + a_8T_8 = (1,1,1)^T$. Since the barycenter lies in the image of $A_6A_7A_8$ we know that each a_i is nonnegative. Since the entries of T_i are also nonnegative we know that each a_i is either 1 or 0 and if a_i is 0 then all the entries of T_i are either 1 or 0. So if $a_6 = a_7 = a_8 = 1$ then after reordering we must have $A_i = A_{5+i}$ and the first possibility (i) holds. If after reordering we have $a_6 = a_7 = 1$ and $a_8 = 0$ then after more reordering we must have $T_6 = (1,1,0)^T$ and $T_7 = (0,0,1)^T$ and possibility (iii) holds. Finally, if $a_6 = 1$ and $a_7 = a_8 = 0$ then $T_6 = (1,1,1)^T$ and possibility (ii) holds. □

At the moment it is just a hope that the preceding facts will allow you to understand the dimension three case. But they do give powerful information. In particular, let us see what Lemma 7.4 tells us in the three dimensional case. Note that $\mathcal{K}(X, \mathcal{A})$ is two dimensional. Now look at a 2-simplex of $\mathcal{K}(Y, \mathcal{B})$ and apply Lemma 7.4. If the first case holds, we are happy. In the second case, note that the exponent of A_4 in B_i must be 1. To see this, note that these exponents are all equal since we map to the barycenter. But if they have a nontrivial common divisor, it violates the determinant ± 1 condition of Theorem 7.7. It is natural to hope that blowing up $B_1 \cap B_2 \cap B_3$ will improve matters. It may be necessary to blow up X some so that f lifts though, so this is a delicate issue. Good invariants must be found. If the last case holds, we can then blow up Y along the curve $B_2 \cap B_3$ and then f lifts, sort of. Unfortunately, there may be other 2-simplices with the edge B_2B_3 and you must consider them also, but anyway you have hope. I presume that something useful comes out of this anyway, but it is definitely work in progress which I invite others to investigate. I have a computer program which helps compute examples and may be of use, especially on some platforms where one can see a picture of the subdivisions being done. In computer experiments with random examples, I have an algorithm for choosing the center to blow up and this algorithm appears to work well, converging to the desired situation where $\mathcal{K}(f)$ takes vertices to vertices. I have only tested it on random examples with a couple hundred vertices, since in intermediate stages there tend to be several thousand vertices which taxes the limits of my computer, but these results are quite encouraging even though it is not yet clear how to prove the algorithm will always converge.

Finally, since I refer to Proposition 3.3.9 of [**1**] and its proof is somewhat complex, I give a proof which works in the constant exponent case which is needed here. It is based on the alternative proof sketched at the end of the proof in [**1**]. The simplicial complex point of view allows some simplifications to be made.

THEOREM 7.9. *Let $f\colon (X, \mathcal{A}) \to (Y, \mathcal{B})$ be an exact tico map with constant exponents and let $\pi\colon (Y', \mathcal{B}') \to (Y, \mathcal{B})$ be a tico blowup whose center C is an intersection of sheets. Then there is a sequence of tico blowups $\rho\colon (X', \mathcal{A}') \to (X, \mathcal{A})$ and a tico map $f'\colon (X', \mathcal{A}') \to (Y', \mathcal{B}')$ so that*

(i) $\pi \circ f' = f \circ \rho$.

(ii) *All the centers of the blowups in ρ are the intersections of exactly two sheets.*

PROOF. Let $B_1, \ldots, B_k$ be the sheets of $\mathcal{B}$ so that $C = \bigcap_{i=1}^k B_i$. Let σ be any simplex of $\mathcal{K}(Y, \mathcal{B})$ which contains all the vertices $B_1, \ldots, B_k$. Label its vertices $B_1, \ldots, B_m$. Let $F \subset \sigma$ be the set of points with barycentric coordinates $(t_1 : \ldots : t_m)$ so that $t_1 = t_2$ and $t_1 \leq t_i$ for all $i \leq k$. Then F has codimension one in σ and is the image of a simplex in $\mathcal{K}(Y', \mathcal{B}')$. We will find a sequence of blowups ρ so that for each simplex δ of $\mathcal{K}(X', \mathcal{A}')$, the interior of δ is either disjoint from or contained in $\mathcal{K}(f\rho)^{-1}(F)$. It suffices to do this, since then we may do it for each such simplex σ and each ordering of $B_1, \ldots, B_k$, after which Theorem 7.5 shows that the lifting f' exists.

If A_1A_2 is a 1-simplex of $\mathcal{K}(X, \mathcal{A})$ define a nonnegative integer $\gamma(A_1A_2)$ as follows. Let $F' \supset F$ be the set of points in σ with barycentric coordinates $(t_1 : \ldots : t_m)$ so that $t_1 = t_2$. Let α_{ij} be the exponent of A_i in B_j. We set $\gamma(A_1A_2) = 0$ if the interior of A_1A_2 does not intersect $\mathcal{K}(f)^{-1}(F')$ or if $\mathcal{K}(f)(A_1A_2) \subset F$. Otherwise, after perhaps reordering we must have $\alpha_{11} > \alpha_{12}$ and $\alpha_{21} < \alpha_{22}$. We then let $\gamma(A_1A_2) = \alpha_{11} - \alpha_{12} + \alpha_{22} - \alpha_{21}$. If δ is any simplex of $\mathcal{K}(X, \mathcal{A})$, we let $\gamma(\delta)$ be the maximum of $\gamma(\delta_1)$ over all 1-simplices δ_1 in δ and we let $\lambda(\delta)$ be the number of 1-simplices δ_1 in δ with $\gamma(\delta_1) = \gamma(\delta)$.

The proof now proceeds as follows. Among all simplices δ of $\mathcal{K}(X, \mathcal{A})$, pick one with lexicographically maximum $(\gamma(\delta), \lambda(\delta))$. If this maximum has $\gamma(\delta) = 0$ then we are done. If it is not 0, we may pick vertices A_1 and A_2 of δ so that $\gamma(A_1A_2)$ is this maximum γ. We then blow up $A_1 \cap A_2$. After blowing up, either the maximum (γ, λ) has decreased, or there are fewer simplices δ with the maximum (γ, λ). Consequently, after a finite number of such steps we are done. $\square$

REFERENCES

1. S. Akbulut and H. King, *The topology of real algebraic sets*, MSRI Pub. 25, Springer-Verlag (1992).
2. E. Bierstone and P. Milman, *A simple constructive proof of canonical resolution of singularities*, Effective Methods in Algebraic Geometry, Progress in Mathematics #94, Birkhaüser (1991) 11-30.
3. H. Hironaka, *Resolution of singularities of an algebraic variety over a field of characteristic zero*, Annals of Math. 79 (1964) 109-326.

DEPARTMENT OF MATHEMATICS, UNIVERSITY OF MARYLAND, COLLEGE PARK, MARYLAND 20742

E-mail address: hking@math.umd.edu

Contemporary Mathematics
Volume **182**, 1995

On Algebraic Vector Bundles

S. AKBULUT AND H. KING

ABSTRACT. We discuss elementary properties of strongly algebraic vector bundles, and give a proof that the classifying map of such bundles to Grassmann variety could be algebraically deformed to an algebraic imbedding.

1. Introduction

Let X be a compact nonsingular real algebraic set. Recall that an $\mathbb{R}^k$ vector bundle $E \to X$ is called a strongly algebraic if E is a nonsingular real algebraic set containing X as the zero section. Let us review some of the useful properties of totally algebraic bundles: Perhaps, the most important totally algebraic vector bundle is the universal bundle over the Grassmanian variety of k planes in $\mathbb{R}^n$:

$$\begin{array}{ccl} E(k,n) & = & \{(A,v) \in \mathcal{M}_{\mathbb{R}}(n) \times \mathbb{R}^n \mid A \in G(k,n),\ Av = v\,\} \\ \downarrow & & \\ G(k,n) & = & \{A \in \mathcal{M}_{\mathbb{R}}(n) \mid A^t = A,\ A^2 = A,\ \mathrm{trace}(A) = k\,\} \end{array}$$

Here $\mathcal{M}_{\mathbb{R}}(n)$ denotes $n \times n$ real matrices $\cong \mathbb{R}^{n^2}$. A useful property of a totally algebraic vector bundle $E \to X$ is that its classifying (Gauss) map

$$\rho_E : X \to G(k,n)$$

(for some large n) is an entire rational map, (cf [**1**]). Conversely any entire rational map from a nonsingular real algebraic set X to $G(k,n)$ induces a totally algebraic bundle over X by pulling back $E(k,n)$. Hence a vector bundle $E \to X$ over a nonsingular algebraic set is isomorphic to a totally algebraic vector bundle if and and only if its classifying map is homotopic to an entire rational map.

1991 *Mathematics Subject Classification.* Primary 57R19, 15PO5; Secondary 14C25.
The first author was supported in part by NSF Grant No. DMS-9102518
No version of this paper will be submitted for publication elsewhere.

Another useful property of a totally algebraic bundle $E \to X$ is that generically any smooth manifold $L \subset X$ representing the Euler class, can be approximated by a nonsingular algebraic subset of X. This is because a transverse section $s : X \to E$ corresponds to a map $\sigma : X \to E(k,n)$, where $\sigma(x) = (\rho_E(x), f(x))$ and $f : X \to \mathbb{R}^n$ is a smooth map with $\rho_E(x)f(x) = f(x)$. We can obtain a rational approximation $\tilde{\sigma}$ of σ by first approximating f by a polynomial map F and letting $\tilde{\sigma}(x) = (\rho_E(x), \rho_E(x)F(x))$. Then $\tilde{\sigma}^{-1}(E(k,n))$ approximates L.

The following Theorem 1.1 is due to M. Buchner who produced a proof at this conference by using resolution of singularities. Here we offer a direct elementary proof as a corollary to an observation on imbedding Grassmannianns:

THEOREM 1.1. *Let $E \to X$ be a totally algebraic $\mathbb{R}^k$-bundle over a compact nonsingular algebraic set. Then for some large n we can find an entire rational map $\rho : X \to G(k,n)$ classifying this bundle, which is an imbedding onto its image.*

We noticed that Theorem 1.1 could be proven easily because of the following elementary result. One can imbed the product of two Grassmannians in another via the tensor product. When $k = \ell = 1$, this is the well known Segre imbedding. The reason this helps is that it gives many linearly independent algebraic sections of the normal bundle of one Grassmanian in another. These sections can then be used to approximate a rational map to the Grassmanian by an imbedding.

THEOREM 1.2. *There is an entire rational function*

$$\beta_{k\ell mn} \colon G(k,n) \times G(\ell,m) \to G(k\ell, mn)$$

which is an isomorphism onto its image. This map is given by tensor product. If $k = 1$ then there is a $q \in G(1,n)$ so that $\beta_{1\ell mn}$ restricted to $q \times G(\ell,m)$ is the usual inclusion $G(\ell,m) \to G(\ell,mn)$.

PROOF. As notation let A_{ij} denote entries of a matrix A. Let A be an $n \times n$ matrix and let B be an $m \times m$ matrix. Then we let $A \otimes B$ denote the $mn \times mn$ matrix whose entry in the $(im+j)$-th row and $(sm+t)$-th column is $A_{is}B_{jt}$ (for convenience, we start indexing with 0 rather than 1).

$$A \otimes B = \begin{pmatrix} A_{00}B & A_{01}B & \cdots & \cdots & \cdots & A_{0,n-1}B \\ A_{10}B & A_{11}B & \cdots & \cdots & \cdots & A_{1,n-1}B \\ \vdots & \vdots & \ddots & & & \vdots \\ \vdots & \vdots & & A_{is}B & & \vdots \\ \vdots & \vdots & & & \ddots & \vdots \\ A_{n-1,0}B & A_{n-1,1}B & \cdots & \cdots & \cdots & A_{n-1,n-1}B \end{pmatrix}.$$

Define $\beta_{k\ell mn}(K, L) = K \otimes L$. Let $V \subset G(k\ell, mn)$ be the subvariety given by the points $M \in G(k\ell, mn)$ satisfying the equations:

$$M_{im+j,sm+t}M_{i'm+j',s'm+t'} = M_{i'm+j,s'm+t}M_{im+j',sm+t'}$$

$$\ell \sum_{t=0}^{m-1} M_{im+j,im+t}M_{im+t,im+t'} = \left(\sum_{t=0}^{m-1} M_{im+t,im+t}\right) M_{im+j,im+t'}$$

for all $0 \le j, j', t, t' < m, \quad 0 \le i, i', s, s' < n$.

In other words, V is the set of matrices M in $G(k\ell, mn)$ which when written in block form $(M(is))$ of $m \times m$ matrices $M(is)$, then all the $M(is)$ are multiples of each other and furthermore, $\ell M(ii)^2 = (trM(ii))M(ii)$ (here tr means trace).

We claim that $\beta_{k\ell mn}$ is an isomorphism to V. First of all, the image of $\beta_{k\ell mn}$ is contained in V since if $M = K \otimes L$ then M is symmetric, $trM = (trK)(trL)$, $M^2 = K^2 \otimes L^2 = K \otimes L = M$. Furthermore each $M(is)$ is a multiple of of L, in particular $M(ii) = K_{ii}L$ hence:

$$\ell M(ii)^2 = \ell K_{ii}^2 L^2 = (\ell K_{ii})K_{ii}L = (trM(ii))M(ii)$$

Now let us see that $\beta_{k\ell mn}$ is onto V and compute its inverse and see that it is rational. To see this, take any matrix M in V. Since $trM = k\ell \neq 0$ there is some i'' and j'' so that $M_{i''m+j'',i''m+j''} \neq 0$. Then

$$M_{i'm+j',s'm+t'} = M_{i'm+j'',s'm+j''}M_{i''m+j',i''m+t'}/M_{i''m+j'',i''m+j''}$$

Let L' be the $m \times m$ matrix with (i,j)-th entry $M_{i''m+i,i''m+j}/M_{i''m+j'',i''m+j''}$ and let K' be the $n \times n$ matrix with (i,j)-th entry $M_{im+j'',jm+j''}$. Then

$$M = K' \otimes L'$$

Note that $k\ell = trM = trK'trL'$ so K' and L' both have nonzero trace. Let $L = \ell L'/trL'$ and $K = kK'/trK'$, then we have $M = K \otimes L$. Note that K and L are both rational functions of M. We claim that $K \in G(k, n)$ and $L \in G(\ell, m)$. Note that K' and L' are symmetric, so K and L are symmetric also. But the second equations for V tell us that $\ell(K_{i''i''}L)^2 = (trK_{i''i''}L)K_{i''i''}L$, so $\ell K_{i''i''}^2 L^2 = \ell K_{i''i''}^2 L$ and thus $L^2 = L$. But we have $M^2 = M$ so $K \otimes L = (K \otimes L)^2 = K^2 \otimes L^2 = K^2 \otimes L$. Thus we have $K = K^2$ since $L \neq 0$. So $\beta_{k\ell mn}$ has a right inverse:

$$\alpha \colon V \to G(k, n) \times G(\ell, m)$$

But α is a left inverse also since if $\beta_{k\ell mn}(K, L) = \beta_{k\ell mn}(K', L')$ we have $K \otimes L = K' \otimes L'$ so there is a number b so that $K' = bK$ and $L' = (1/b)L$. But $k = trK' = btrK = bk$ so $b = 1$.

Notice if $\ell = 1$ we may take q to be the $n \times n$ matrix which is zero everywhere except the entry in the zeroeth row and zeroeth column where it is 1. Then $q \otimes L$ is the matrix which is zero everywhere except the first $m \times m$ block where it is L. This is the usual inclusion $G(\ell, m) \subset G(\ell, mn)$ $\square$

PROOF. (of Theorem 1.1) Let $\pi : E \to X$ be a strongly algebraic vector bundle, and $\rho \colon X \to G(\ell, m)$ be an entire rational map classifying this bundle. Now let $\gamma \colon X \to G(1, n)$ be a regular map whose image is a projectively closed algebraic subset, i.e., its image is a nonsingular algebraic variety and is contained in an affine chart (see [**1**], Theorem 2.5.13). Consider the map $\sigma \colon X \to G(k, mn)$ given by

$$\sigma(x) = \beta_{1\ell mn}(\gamma(x), \rho(x))$$

Then σ also classifies the bundle since it is homotopic to the map:

$$x \mapsto \beta_{1\ell mn}(q, \rho(x)) = \iota\rho(x)$$

where $\iota \colon G(\ell, m) \to G(\ell, mn)$ is the usual inclusion. Clearly σ is an algebraic imbedding onto its image □

REFERENCES

1. Akbulut and H.King, *Topology of real algebraic sets*, MSRI book series no. 25, Springer-Verlag (1992).
2. ———, *The topology of real algebraic sets with isolated singularities*, Annals of Math. 113 (1981) 425-446.
3. Benedetti and R.Tognoli, *On real algebraic vector bundles,*, Bull. des Sci. Math. 2 serie' 104 (1980) 89-112.

DEPARTMENT OF MATHEMATICS, MICHIGAN STATE UNIVERSITY, E. LANSING, MI, 48824
E-mail address: akbulut@math.msu.edu

DEPARTMENT OF MATHEMATICS, UNIVERSITY OF MARYLAND, COLLEGE PARK, MD, 20782
E-mail address: hck@frances.umd.edu

Recent Titles in This Series

(Continued from the front of this publication)

(See the AMS catalog for earlier titles)